Martian Outpost

The Challenges of Establishing a Human ⁣Settlement on Mars

Erik Seedhouse

Martian Outpost

The Challenges of Establishing a Human Settlement on Mars

 Springer

Published in association with
Praxis Publishing
Chichester, UK

Dr Erik Seedhouse, F.B.I.S., As.M.A.
Milton
Ontario
Canada

SPRINGER–PRAXIS BOOKS IN SPACE EXPLORATION
SUBJECT *ADVISORY EDITOR*: John Mason M.B.E., B.Sc., M.Sc., Ph.D.

ISBN 978-0-387-98190-1 Springer Berlin Heidelberg New York

Springer is a part of Springer Science + Business Media (*springer.com*)

Library of Congress Control Number: 2009921645

Cover design: Jim Wilkie
Copy editor: Dr John Mason
Typesetting: BookEns Ltd, Royston, Herts., UK

Printed in Germany on acid-free paper

Contents

Preface

"I am convinced that the future development of the possibilities of your own people, as well as those of mankind, will depend on some of you young people striking boldly out along new tracks."

Legendary explorer, Fridtjof Nansen.
Rectorial address at St. Andrews University, 1926.

Fridtjof Nansen (1861–1930) was perhaps the greatest explorer of all time, a man who led the way in surmounting obstacles and ignoring regular conventions, as exemplified by his 1888 expedition to cross the Greenland ice-cap from east to west, without dogs or sledges.

Like all of Nansen's expeditions, the Greenland project was simple and bold but required thorough preparation and planning. At the outset, the body of expert opinion condemned Nansen's plan as madness, arguing that without dogs and sledges it would be impossible to traverse the inland ice. Nevertheless, Nansen embarked upon his expedition and, after facing delays due to ice and weather, his expedition came ashore on the east coast of Greenland in August 1888. The journey across the ice cap was hazardous and exhausting in the extreme. Nansen's party had to climb nearly three thousand meters above sea-level, navigate through huge fissures in the ice and suffer temperatures that regularly plummeted below minus forty degrees Celsius. Despite the adversities, Nansen's team completed the descent to the west coast and through their expedition, made a decisive contribution to the scientific knowledge of the interior of Greenland, which many experts had mistakenly believed to be free of ice.

Nansen's Greenland crossing, however, was completely overshadowed by his next expedition to the polar region. Once again, following a simple, elegant yet unorthodox plan, Nansen's idea was to sail a ship as far east as possible along the Siberian coast, and then allow the vessel to be frozen in the ice, in the hope it would be carried across the North Pole or a point close to it. Once again, experts dismissed Nansen's plan as folly but, choosing to ignore the opinion of the naysayers, Nansen set out in the summer of 1893 in the *Fram*, a ship specifically designed to withstand the pressure of the polar ice. In September, the *Fram* was trapped in the ice and began drifting in a north-westerly direction, once again confirming Nansen's theory.

After a year trapped in the ice, Nansen had to accept the *Fram* would not be carried as far north as he had hoped. Instead, bold as ever, Nansen decided to leave the ship with companion, Hjalmar Johansen, and continue northwards across the pack ice, using skis and sledges drawn by dogs. Taking supplies sufficient for only one hundred days, Nansen and Johansen set out in March 1895 as *Fram* continued on her course. Struggling across frozen seas and almost impassable masses of ice towards an unknown fate, the intrepid explorers endured a journey of incredible privation and danger, eventually reaching a point that was the closest a human had ever been to the North Pole before being forced to turn back due to lack of food. After 132 days, Nansen and Johansen sought refuge on Franz Josef Land, a bleak and desolate island, where they spent the winter in a primitive shelter built of rocks and ice. The following year, the two explorers continued their trek south, eventually encountering British explorer, Frederick Jackson, who took them back to Norway on his ship. The *Fram* returned to Norway at almost the same time, having drifted for three years, as predicted by Nansen.

The adventures of the *Fram* and her crew attracted worldwide attention and established Nansen's reputation as one of the great pioneers of polar exploration. The scientific results realized by the expedition were no less impressive than the courage of the explorers, since the data collected proved of great value in the disciplines of arctic meteorology, oceanography and zoology.

When Nansen set out on his first great expedition over Greenland, there were still large blank areas on the maps of the world. By the time he died in 1930, there were hardly any left, a fact that saddened the great Norwegian who was driven by exploring the unknown. More than one hundred years after Nansen's epic expeditions, uncharted lands still await discovery, but this time we must venture a little further than the eminent Norwegian explorer. Traveling to Mars will provide a challenge on the same scale as Nansen's expeditions to Greenland and the high arctic. Such an expedition will be one fraught with danger and risk, but pioneering such a venture will be no less noble a challenge than the expeditions embarked upon by Nansen and his fellow explorers more than a century ago.

Civilizations thrive on challenge and decay without it and the time is long past for humans to once again face outward and embrace the bold endeavor of travelling to Mars. In so doing, we will make a profound statement testifying to the enduring human spirit. We must not shrink back from this task but attack it with the same enthusiasm and intelligence that Nansen brought to the challenges of his time.

The heroes of one generation are too easily forgotten by the next and the world is in dire need of rediscovering the bold example of Fridtjof Nansen. A manned mission to Mars is the key to reviving the spirit of exploration and once again expanding the limits of the possible. This book describes how we will embark on such a bold endeavor.

Acknowledgments

In writing this book, I have been fortunate to have had my wife, Doina Nugent as my proof-reader. Once again, she has applied her considerable skills to make the text as smooth and coherent as possible.

I would like to thank Bob of John Frassanito and Associates for giving me permission to use some of the outstanding images that appear in this book. I am also grateful to SpaceWorks Engineering Inc.'s Mark Elwood, Director, Space Media Group, and Brad St. Germain, Director of Advanced Concepts for permission to use the images illustrating SEI's Mars mission architecture. Thanks also to space artist, Adrian Mann, who created the wonderful concept art of the Skylon spaceplane and provided permission to use his images and to Guillermo Trotti of Trotti and Associates for permission to use their art illustrating extreme expeditionary architecture. Many thanks also to António H.F. Maia of the DIRECT team for permission to use the wonderful images illustrating the Jupiter launch system.

I am also grateful to the five reviewers who made such positive comments concerning the content of this volume and to the Publisher, Clive Horwood, and his team at Praxis for guiding this book through the publication process. The author also gratefully acknowledges the Chief Subject Advisory Editor for Praxis, Dr John Mason, whose attention to detail and patience greatly facilitated the publication of this book.

Finally, this book is dedicated to the countless individuals who have worked on the designs of future interplanetary spacecraft, many of which will never be developed, and to those engineers and scientists working on developing the Orion and Ares family of launch vehicles that will one day transport astronauts millions of kilometers to the Red Planet.

To
Rose-Marie, Richard and Tracy

About the author

Erik Seedhouse is an aerospace scientist and manned spaceflight consultant with ambitions to become an astronaut. After experiencing his first taste of microgravity during the European Space Agency's 22nd Parabolic Flight Campaign in 1995 he gained his Ph.D. while working at the German Space Agency's Institute for Space Medicine in Cologne. Recently, Erik worked as an astronaut training consultant for Bigelow Aerospace in Las Vegas and wrote 'Tourists in Space', the training manual for spaceflight participants. His company – Spaceflight Solutions – provides training programs and advice to the private spaceflight sector. He is a Fellow of the British Interplanetary Society and a member of the Aerospace Medical Association. An avid pilot and scuba diver, Erik also races Ultraman triathlons, climbs mountains and spends as much time as possible in Kona on the Big Island of Hawaii and at home in Sandefjord, Norway. Erik lives with his wife and two cats on the Niagara Escarpment in Canada.

Figures

(* Figures marked with an asterisk also appear in the color section positioned between pages 160 and 161)

Tables

Abbreviations and acronyms

AAA	Avionics Air Assembly
ACOS	Attitude and Orbit Control System
ACT	Advanced Concepts Team
AI	Artificial Intelligence
ALARA	As Low As Reasonably Achievable
ALHAT	Autonomous Landing and Hazard Avoidance Technology
ALS	Advanced Life Support
AMPDXA	Advanced Multiple Projection Dual Energy X-ray Absorptiometry
AOD	Automatic Opening Device
AQF	Astronaut Quarantine Facility
ARC	Ames Research Center
ARS	Acute Radiation Syndrome
ASCR	Astronaut Strength Conditioning and Rehabilitation
ASG	Astronaut Spouses Group
ASI	Agenzia Spaziale Italiana
ASRG	Advanced Stirling Radioisotope Generator
ATCO	Ambient Temperature Catalytic Oxidation
ATCS	Active Thermal Control System
ATHLETE	All Terrain Hex-Limbed Extra Terrestrial Explorer
ATSS	Advanced Transportation System Studies
ATV	All Terrain Vehicle
BEES	Bioinspired Engineering of Exploration Systems
BMD	Bone Mineral Density
BMI	Bismaleimide
BNL	Brookhaven National Laboratory
BNSC	British National Space Centre
BNTR	Bimodal Nuclear Thermal Rocket
BPC	Boost Protective Cover
BPS	Biomass Production System
CaLV	Cargo Launch Vehicle
CAPCOM	Capsule Communicator
CARD	Constellation Architecture System Requirements

C&C	Command and Control
C&N	Communications and Navigation
CCAA	Common Cabin Air Assembly
CCB	Common Core Booster
CCDH	Command and Control Data Handling
CEV	Crew Exploration Vehicle
CFD	Computational Fluid Dynamics
CGI	Computer Generated Image
CLV	Cargo Launch Vehicle
CM	Crew Module
CME	Coronal Mass Ejection
CMO	Crew Medical Officer
CMRS	Carbon Moisture Removal System
CNES	Centre Nationale d'Etudes Spatiales
CNM	Computational Network Modeling
CNS	Central Nervous System
CNSA	Chinese National Space Administration
CONUS	Continental United States
COSPAR	Committee on Space Research
CPT	Constant Power Throttling
CRV	Crew Return Vehicle
CSA	Canadian Space Agency
CSAV	Crew Surface Ascent Vehicle
CSIRO	Commonwealth, Scientific Industrial Research Organization
CT	Computer Tomography
DAEZ	Downrange Abort Exclusion Zone
DARPA	Defense Advanced Research Projects Agency
DDOR	Delta Differential One-way Ranging
DDT&E	Design, Development, Test and Evaluation
DEXA	Dual Energy X-ray Absorptiometry
DLR	Deutsche Zentrum für Luft und Raumfahrt
DRM	Design Reference Mission
DS	Descent Stage
DSENDS	Dynamic Simulator for Entry, Descent and Landing
DTE	Direct to Earth
EAC	European Astronaut Centre
EAGLE	Evolved Acceleration Guidance Logic for Entry
ECLSS	Environmental Control Life Support System
EDL	Entry, Descent and Landing
EDS	Earth Departure Stage
EELV	Evolved Expendable Launch Vehicle
EI	Entry Interface
EMS	Electronic Monitoring System
EMU	Extravehicular Mobility Unit
EPS	Electrical Power System

ERS	Earth Return Stage
ERV	Earth Return Vehicle
ESA	European Space Agency
ESAS	Exploration Systems Architecture Study
ESM	Equivalent System Mass
ESMD	Exploration Systems Mission Directorate
ET	External Tank
EU	European Union
EVA	Extravehicular Activity
EXEMSI	Experimental Campaign for European Manned Space Infrastructure
FBR	Fixed Base Radio
FDA	Federal Drug Administration
FDIR	Failure Detection Isolation and Recovery
FSO	Family Support Officer
FSP	Fission Power System
FSRCS	First Stage Roll Control System
FTV	Flight Test Vehicle
GAC	Global Aerospace Corporation
GCR	Galactic Cosmic Radiation
GES	Global Exploration Strategy
GHe	Gaseous Helium
GOX	Gaseous Oxygen
GPC	General Purpose Computer
GPS	Global Positioning System
GRC	Glenn Research Center
GSFC	Goddard Space Flight Center
HDPE	High Density Polyethylene
HEPA	High Efficiency Particulate -
HGDS	Hazardous Gas Detection System
HMD	Head Mounted Device
HMP	Haughton Mars Project
HPOGA	High Pressure Oxygen Generator Assembly
HUBES	Human Behavior in Extended Spaceflight
HUT	Hard Upper Torso
HZE	High Energy Particles
IARS	Irradiation-Acute Radiation Syndrome
IBMP	Institute for Biomedical Problems
ICAN	Ion Compressed Antimatter Nuclear
ICRP	International Commission on Radiological Protection
IMLEO	Initial Mass in Low Earth Orbit
IMU	Inertial Measurement Unit
INS	Inertial Navigation System
IR	Infrared
ISDC	International Space Development Conference
ISEMSI	Isolation Study for European Manned Space Infrastructure

ISP	In Space Propulsion
ISPS	In Space Propulsion Stage
ISRO	Indian Space Research Organization
ISRU	In Situ Resource Utilization
ISS	International Space Station
ITH	Inflatable Transfer Habitat
IUA	Instrument Avionics Unit
JAXA	Japan Aerospace Exploration Agency
JCC	Jupiter Common Core
JLS	Jupiter Launch System
JPL	Jet Propulsion Laboratory
JSC	Johnson Space Center
KARI	Korea Aerospace Research Institute
KSC	Kennedy Space Center
LADAR	Laser Detection and Ranging
LANTR	Liquid Oxygen Augmented Nuclear Thermal Rocket
LAS	Launch Abort System
LBNP	Lower Body Negative Pressure
LCD	Liquid Crystal Display
LEO	Low Earth Orbit
LES	Launch Escape System
LET	Linear Energy Transfer
LH2	Liquid Hydrogen
LLO	Low Lunar Orbit
LLTV	Lunar Landing Training Vehicle
LMO	Low Mars Orbit
LOC	Loss of Crew
LOM	Loss of Mission
LOX	Liquid Oxygen
LRC	Langley Research Center
LSAM	Lunar Surface Access Module
LSS	Life Support System
MAF	Michoud Assembly Facility
MARSAT	Mars Aerostationary Relay Satellite
MCP	Mechanical Counter-Pressure
MCS	Mars Cycler System
MCT	Mars Communication Terminal
MDRS	Mars Desert Research Station
MDS	Mars Departure Stage
MDU	Manufacturing Demonstration Unit
MEV	Mars Excursion Vehicle
MEX	Mobile Exploration System
MIT	Massachusetts Institute of Technology
MLAN	Martian Local Area Network
MLI	Multi Layer Insulation

MOI	Mars Orbit Insertion
MPD	Magnetoplasmadynamic Thruster
MPSS	Main Parachute Support Structure
MRI	Magnetic Resonance Imaging
MSFC	Marshall Space Flight Center
MSL	Mars Science Laboratory
MSH	Mars Surface Habitat
MSR	Mother Ship Rover
MSRM	Mars Sample Return Mission
MTF	Magnetized Target Fusion
MTS	Mars Transfer Stage
MTSV	Mars Transfer and Surface Vehicle
NBF	Neutral Buoyancy Facility
NCRP	National Commission on Radiation Protection
NEEMO	NASA Extreme Environment Mission Operations
NEO	Near Earth Object
NERVA	Nuclear Engine for Rocket Vehicle Applications
NIAC	NASA Institute for Advanced Concepts
NOAA	National Oceanic Atmospheric Administration
NPO	Nuclear Propulsion Office
NSAU	National Space Agency of Ukraine
NSF	National Science Foundation
NTP	Nuclear Thermal Propulsion
NTR	Nuclear Thermal Rocket
OBS	Operational Bioinstrumentation System
OML	Outer Mold Line
OMS	Orbital Maneuvering System
PAO	Public Affairs Officer
PBAN	Polybutadiene Acrylonitrite
PCAD	Propulsion and Cryogenics Advanced Development
PCU	Power Control Unit
PDF	Probability Distribution Function
PDV	Powered Descent Vehicle
PET	Positron Emission Tomography
PICA	Phenolic Impregnated Ceramic Ablator
PKA	Protein Kinase A
POST	Program to Optimize Simulated Trajectories
PRC	Peripheral Retropropulsive Configuration
PSG	Psychological Services Group
RBC	Reference Ballute Configuration
RBE	Relative Biological Effectiveness
RCS	Reaction Control System
RCM	Reciprocating Chemical Muscle
RDA	Recommended Daily Allowance
RDT	Rendezvous and Docking Technology

RLV	Reusable Launch Vehicle
ROV	Remotely Operated Vehicle
RSRB	Reusable Solid Rocket Booster
RVE	Rendezvous Experiment
SA	Spacecraft Adapter
SAR	Search and Rescue
SARSAT	Search and Rescue Satellite Aided Tracking
SAS	Space Adaptation Syndrome
SEI	SpaceWorks Engineering Inc.
SEI	Space Exploration Initiative
SEIM	Surface Endoskeletal Inflatable Module
SETI	Search for Extraterrestrial Intelligence
SFU	Simon Fraser University
SM	Service Module
SOHO	Solar and Heliospheric Observatory
SPE	Solar Particle Event
SRL	Space Radiation Laboratory
SRM	Solid Rocket Motor
SRR	Systems Requirement Review
SRRP	Space Radiation Research Program
SRSF	Slow Release Sodium Fluoride
SSC	Systems and Software Consortium
SSH	Storm Shelter Hatch
SSME	Space Shuttle Main Engine
SST	Single Systems Trainer
SSTO	Single Stage to Orbit
STP	Stability Transition Phenomenon
SVD	Synthetic Vision Display
TAI	Trotti and Associates Inc.
TAL	Targeted Abort Landing
TCM	Trajectory Correction Maneuver
TCP/IP	Transmission Control Protocol/Internet Protocol
TDP	Terminal Descent Phase
TDS	Terminal Descent Sensors
TEI	Trans Earth Insertion
TLI	Trans Lunar Insertion
TMI	Trans Mars Insertion
TO	Thrust Oscillation
TPS	Thermal Protection System
UAS	Untargeted Abort Splashdown
UAV	Uninhabited Aerial Vehicle
UHF	Ultrahigh Frequency
ULWB	Ultra Lightweight Ballute
UV	Ultraviolet
VASIMR	Variable Specific Impulse Magnetoplasma Rocket

VCC	Voice Call Continuity
VDC	Volts Direct Current
VEG	Virtual Environment Generator
VR	Virtual Reality
VSE	Vision of Space Exploration
WBC	White Blood Cell
WET-F	Weightlessness Environment Training Facility
WFF	Wallops Flight Facility
WRS	Waster Removal System

1

Why go?

"I have decided today that the United States should proceed at once with the development of systems and technologies designed to take American astronauts on landing missions to Mars. This system will center on a new generation of rockets, exploiting nuclear power, which will revolutionize and render routine long-haul interplanetary flights.

The year 1971 was a year of conclusion for America's current series of manned flights to the Moon. Much was achieved in the three successful landing missions – in fact, the scientific results of the third mission have been shown to greatly outweigh the return from all earlier manned spaceflights, to Earth orbit or the Moon. But it also brought us to an important decisions point – a point of assessing what our space horizons are as Apollo ends, and of determining where we go from here.

In the scientific arena, the past decade of experience has taught use that spacecraft are an irreplaceable tool for learning about our near-space environment, the Moon, and the planets, besides being an important aid to our studies of the Sun and stars. In utilizing space to successfully meet needs on Earth, we have seen the tremendous potential of satellites for international communications and worldwide forecasting, and global resource monitoring.

However, all these possibilities, and countless others with direct and dramatic bearing on human betterment, will not be achieved without a continuation of the dream which has carried us so far and so fast: I mean the dream of exploration, of American and human expansion into space, the greatest frontier of all. In my decision today, I have taken account of the need to fully encourage and sustain that dream.

NASA and many aerospace companies have carried out extensive design studies for the Mars missions. Congress has reviewed and approved this effort. Preparation is now sufficient for us to confidently commence a new development program. In order to completely minimize technical and economic risks, the space agency will continue to cautiously take an evolutionary approach in the development of this new system. Even so, by moving ahead at this time, we can have the first components of the Mars spacecraft in manned flight test by the end of this decade, and operational a short time later. But we will not set arbitrary deadlines, as some have called for; we will make decisions as to the pace of our program in the fullness of time and with the wisdom of experience.

It is for this reason of technological robustness that I have decided against the development of the reusable Space Shuttle at this time; despite the manifest economic benefits of such a launch system if available, I am not convinced that our technology is so mature that we are ready yet to confidently tackle the huge problems posed by the project without cost overruns and delays, and many of its economic benefits should in any case be realizable from enhancements to our existing 'throwaway' platforms.

It is also significant that this major new national enterprise will engage the best efforts of thousands of highly skilled workers and hundreds of contractor firms over the next several years. The continued preeminence of America and American industry in the aerospace field will be an important part of the Mars mission's payload.

We will go to Mars because it is the one place other than our Earth where we expect human life to be sustainable, and where our colonies could flourish. We will go to Mars because an examination of its geology and history will reflect back on a greatly deepened understanding of our own precious Earth.

Above all, we will go to Mars because it will inspire us to clearly look beyond the difficulties and divisions of today, to a better future tomorrow.

'We must sail sometimes with the wind and sometimes against it,' said Oliver Wendell Holmes, 'but we must sail, and not drift, nor lie at anchor.' So with man's epic voyage into space – a voyage the United States of America has led and still shall lead. Apollo has returned to harbor. Now it is time to swiftly build new ships, and to purposefully sail farther than our ancestors could ever have dreamed possible."

President Richard Nixon

Just eleven months after he made this speech, President Nixon cancelled the remaining four Apollo flights and told the Apollo 17 astronauts they might be the last people to walk on the Moon in the twentieth century. It was a dark day for manned space exploration and since Nixon's speech, there have been few presidential initiatives announced and even fewer realized. For example, in 1982, a United States Congress report announced NASA was planning 487 Space Shuttle flights between 1982 and 1994, a launch rate of one every ten days. The report boldly claimed there would be seven Shuttle missions in 1982, eighteen in 1983, twenty-four in 1984, thirty-four in 1985 and forty-five in 1986. Then in 1984, on the evening of January 25th, President Reagan declared in his State of the Union Address that he was directing NASA to develop a permanently manned space station and to accomplish this objective within a decade. Meanwhile, the pace of Shuttle flights was not keeping pace with the Congress report two years earlier. In 1984 there were five Shuttle missions, followed by nine in 1985, and fifteen were planned for 1986, but on the tragic day of January 28, 1986, the Space Shuttle Challenger fell from the sky, killing seven astronauts and with it also terminating NASA's manned space program. A week later, President Reagan reiterated his commitment to the space station project, a goal that was repeated by his successor, President Bush, who, on July 20, 1989, announced his Space Exploration Initiative (SEI) to not only build the space station but also to return humans to the Moon so they might venture on to Mars. With so many announcements and so little substance, the 1980s was a decade in which grandiose visions of manned spaceflight existed as paper plans and little else.

In the early 1990s, the SEI was quietly cancelled and plans for returning to Mars were put on indefinite hold. Meanwhile, Russia continued to operate Mir, its space station, inviting astronauts from the United Kingdom, France, Japan, Germany and finally, in June 1995, the United States, when the Space Shuttle Atlantis performed the first rendezvous and docking with the Russian space station during STS-71. The visit of Atlantis to Mir marked the beginning of a new era in manned spaceflight and acted as the catalyst for the building of the International Space Station (ISS), but the bright future of manned missions to Mars announced in the 1970s and 1980s was yet to be realized.

Worse was yet to come, when, on February 1, 2003, the Space Shuttle Columbia disintegrated high above Texas, a tragedy that was to change not only the face of manned spaceflight but, ultimately and ironically, set NASA on a trajectory for Mars. Less than a year later, on January 14, 2004, President George W. Bush made his Vision for Space Exploration (VSE) announcement, stating NASA would complete the ISS by 2010, before moving on to the Moon and eventually Mars.

The human exploration of Mars will be a venture confirming the ability of humans to leave their home planet and take a significant step into the Solar System. While it may be a small step on a cosmic scale, a manned expedition to Mars will represent a milestone in human achievement because it will require leaving Earth with extremely limited return capability and very few abort options. Once the propulsive maneuver has been performed sending the astronauts on their way to Mars, the crew will be unable to return until the alignment of Mars and Earth permit their return, a return window that may be three years distant.

Although due to the time-lag between Mars and mission control such a mission will mark the first time astronauts will be required to solve their own emergencies, the challenges faced by the crew will be no more hazardous than those faced by polar explorer Fridtjof Nansen and his colleagues more than a century before (Table 1.1 and 1.2).

Given the risks associated with traveling to Mars, many people question why we should embark upon such a mission. Although several articles and even entire books have been written on the subject of why humans should travel to the Red Planet, the answer is really quite simple. Human *curiosity* compels us to explore and, perhaps more importantly, to *understand*. In fact, space exploration, it can be argued, is today's expression of human curiosity and a resulting imperative to explore the unknown, a drive that triggers human inspiration and innovation, irrespective of the risks. That said, a manned mission to Mars will be the most ambitious and risky space mission to date and it is important to understand why such a mission should be undertaken and what the benefits will be.

Table 1.1 Risks of manned mission to Mars

Event	Description
Launch	Equipment or systems failure. Thrust oscillation can tear apart the vehicle. Explosive decompression can kill the crew.
Orbital Operations	Collision with transit vehicle during rendezvous and docking maneuver may result in rapid/explosive decompression.
Transit to Mars	Radiation from unexpected flare may cause radiation sickness leading to incapacitation or death. Failure of communications capability would require complete autonomy of crew operation. Rapid/explosive decompression caused by micrometeorite strike could incapacitate or kill crew.
Arrive in Mars orbit	Aerocapture into low Mars orbit must be conducted flawlessly when crew is in a deconditioned state. Collision with descent/ascent vehicle during rendezvous and docking maneuver may result in rapid/explosive decompression.
Entry, descent and landing	Parachute decelerating the vehicle could fail. Supersonic retro-propulsion could make vehicle unstable. Aeroshell could fail to separate. Guidance and/or inertial navigation units could fail. Vehicle could reach subsonic regime too close to the surface.
Surface operations	Radiation storms could incapacitate crew. Rapid or explosive decompression of habitat could kill crew.
Ascent to orbit	Ascent vehicle could fail leaving crew stranded on Mars. Rendezvous and docking with transit vehicle may be unsuccessful forcing crew to abort to the surface. Transit vehicle systems may malfunction due to cold soak in space.
Trans-Earth insertion	Thrusters could fail after prolonged cold soak in space. Fuel may have boiled off during extended time in space, leaving astronauts stranded in orbit.
Entering Earth orbit	Trajectory alignment may result in vehicle missing entry corridor to orbit.
Descent to Earth surface	Vehicle may de-orbit incorrectly, resulting in a ballistic re-entry, subjecting de-conditioned crew to excessive G forces. Parachutes may fail to deploy.

Table 1.2 Risks of polar exploration

Factor	Description
Sea Conditions	Large sheets of floating ice constantly threaten to crush ships.
Environmental Hazards. Weather	Extreme cold resulting in frostbite and in some cases, amputation. Risk of snow blindness, hypothermia and cold-related changes in peripheral circulation. Storms.
Environmental Hazards. Physical	Large ice fissures and crevasses. Open leads resulting in loss of provisions and/or crewmembers. Attacks by polar bears.
Nutrition	Starvation due to loss and/or exhaustion of provisions, leading to scurvy and in some cases, cannibalism.
Physical Exertion	Dehydration. Exhaustion. Suppression of immune system.
Navigation	Many areas uncharted, leading to unplanned over-wintering and expeditions becoming lost, resulting in loss of crew.

BENEFITS OF TRAVELING TO MARS

Science

Mars (Figure 1.1) is an intriguing planet that has always been an attractive subject for science, which is why so many robotic probes have been sent there. While a veritable armada of robotic missions have visited Mars and returned with a wealth of scientific data, many scientific questions still exist about Mars and its history. These questions will remain until humans set foot on the surface of the Red Planet and begin to conduct science experiments beyond the capability of any robot, no matter how well designed.

The Martian atmosphere, consisting mostly of carbon dioxide, is characterized by a tenuous surface pressure typically less than one percent that of Earth's and a surface temperature varying between 20°C and –140°C. At these pressures and temperatures, water cannot exist in liquid form on the surface, yet the Viking Orbiter missions and NASA's Mars Global Surveyor and Mars Reconnaissance Orbiter spacecraft have observed features indicating that liquid water was present on Mars some time ago and this has been confirmed by the Mars Exploration Rovers and Phoenix lander on the surface. These observations led to three questions forming the core of the scientific exploration of Mars, as defined by NASA:

1. What was the reason for the change in the atmospheric conditions on Mars?
2. What are the implications of such changes for environmental changes on Earth?
3. Is it possible that life arose in the early history of Mars and if it arose, can it still be found somewhere on Mars?

Endeavoring to find answers to these questions provides a framework not only for

Figure 1.1 A whole disk image of the planet Mars taken by the Mars Global Surveyor spacecraft in June 2001. It is winter in the southern hemisphere and there is dust storm activity in the Tharsis volcanic region. (Malin Space Science Systems/NASA.) See Plate 1 in color section.

science but also for the human exploration strategy on Mars. Although robotic missions can provide data that may prove helpful in the search for evidence of past life, an unmanned exploration strategy constitutes an imprecise, lengthy and expensive investigative means to answering the myriad questions related to the search for life.

First, the absolute time scale for the development of Martian surface features must be established, requiring geological investigation to determine formation ages of the volcanic plains, the crust and the water-formed channels.

Next, studies must be conducted to establish evidence for the distribution of water on the surface, requiring investigation of localized deposits such as those around hydrothermal vents, in permafrost and the polar ice caps.

If verifiable evidence for the existence of water *is* established, the next science phase will seek to determine the distribution and characteristics of carbon and nitrogen and also to establish morphological evidence linking concentrations of these elements to the past existence of life. Finally, if organic evidence is determined, scientists will want to know how common organics are on Mars and if the biological materials are similar to those found on Earth.

Although the answers to some of these questions may be revealed by deploying robotic probes to conduct orbital mapping and sample return missions, the sheer scale of the science required to adequately answer these questions necessitates human exploration. People may question the cost of several robotic missions versus that of a single manned mission. The projected cost of the Mars Sample Return Mission (MSRM) planned for 2018 is in excess of eight billion dollars. Although the MSRM will serve as an important technology demonstration confirming the ability of space agencies to send a vehicle to Mars and bring it back, eight billion dollars is a high price for just half a kilogram of Mars rocks. Furthermore, even multiple MSRMs will not be capable of answering many of the strategic questions as well as a human mission.

Human expansion

The human exploration of Mars is tied to the belief that a new world can create new opportunities, a principle common in human history when migrations of people have been motivated by exhaustion of resources or competitive advantage. Historically, the settlement of new territories has often been achieved by those few people who were adventurous enough to go it alone and adopt new terrain as their own, a trend that has largely been reversed with the technological revolution of the twentieth and twenty-first centuries. However, this technological revolution provides the opportunity to once again expand human diversity and move beyond the confines of Earth by establishing settlements on Mars.

Establishing a permanent human presence on Mars will not be easy, mainly due to the absence of a life-sustaining environment and the cost of transporting cargo from Earth. The restrictive launch return windows, which open up every twenty-six months, will necessitate developing highly reliable and rugged systems, but the creation of a livable, artificial environment is technically achievable and the cost of transportation will eventually be reduced.

Once transportation costs have been reduced, the strategy to establish a permanent human settlement on Mars will comprise three distinct stages. First, self-sufficiency will need to be demonstrated by utilizing natural resources for energy, agricultural raw materials and building materials. Enabling in-situ resource utilization (ISRU) will require the development of extraction technology and highly advanced life support systems which will recover most of the waste products from human activity. Second, the physiological and psychological requirements for humans to survive on Mars for extended periods must be met. Physiological requirements include the development of advanced life support systems, enhanced medical capabilities and reducing the risks of radiation exposure and bone deconditioning. Psychological needs will be met by research conducted during long-duration missions, evaluation of small communities living in space analogs such as nuclear submarines and Antarctic research stations, and improved crew communications between Earth and Mars. Third, and perhaps most important, the risks faced by those living on the surface of Mars must be compatible with the perceived benefits.

International cooperation

Although many believe manned spaceflight to be the province of the United States and Russia, a manned mission to Mars will probably be an international enterprise. NASA's current roadmap for putting humans on Mars is to build the infrastructure, pave the way, get everything in place and then invite the Europeans and other international space agencies to the off-ramp. Although it is possible the United States may follow such a plan, in reality, given the magnitude of capability, the technological means and the financial resources required to achieve the objective of a manned mission to Mars, a broader international cooperation will probably be required. Not only will such an approach enable such a mission to be achieved within a shorter timeframe, a global program to explore Mars would also unite European states and more established space powers such as the United States and Russia.

Technological advancement

Although no major technological breakthroughs are required to embark upon a manned mission to Mars, to send humans to the Red Planet and return them safely still requires advanced development of several mission systems. For example, the efficiency of propulsion technology will need to be improved, as will the ruggedness of life support capabilities and the degree of automation in vehicle systems and subsystems. By advancing technology in these and other mission-related systems, a manned Mars mission will result in the maturation of a variety of technologies, the outcome of which will be smaller, lighter and more efficient systems, attractive to entrepreneurs. Furthermore, advances in propulsion and life support technologies will reduce the probability of loss of mission (LOM) or loss of crew (LOC) and lead to a reduction of the size of the transportation system required to move people and supplies to Mars.

Human performance

Extended periods of zero gravity exert several deleterious effects upon the human body, ranging from radiation exposure and immune system suppression to muscle atrophy and the insidious effects of bone demineralization. Various countermeasures have been implemented to counter bone demineralization, ranging from long periods of exercise to lower body negative pressure (LBNP) devices, but to date no effective means exists to prevent astronauts from losing bone mass. Consequently, the prospect of a deconditioned astronaut suffering a fracture during a surface excursion on Mars becomes a very real danger and deconditioning therefore, is a critical issue for Mars missions. While the prospect of breaking a bone on arrival may seem daunting to a crewmember preparing to embark on a Mars mission, given the need to reduce the effects of zero gravity upon the human body, it is possible new strategies and methods will be developed so that humans arriving on the Red Planet can do so safely and without long-term consequences.

Inspiration

There will always be those who ask why we must spend billions of dollars to send humans to a distant planet. Those same people argue that we have to solve the problems on Earth before we spend money on space. Such people are in need of a reality check because in five hundred or a thousand years, we are still going to be talking about the problems that need to be solved, and to think that the human race can attain some utopian state where all problems will be solved is quite simply delusional. The real problem right now is not embarking on missions to Mars when we have the capability to do so.

Manned spaceflight is at a turning point and has been since the end of the Apollo missions in 1972. By stagnating in low Earth orbit (LEO) for more than three decades, the incremental future of humans in space has simply stalled. Meanwhile, on Earth, humanity has devolved into a rudderless society, raising children in a world characterized by mind-numbing affluence, point-and-shoot video games and morally bankrupt leaders. Devoid of heroes, the world in which children grow up is also one in which the scientific quest for understanding and exploration is just not cool. In such dire circumstances it is not a case of an expedition to Mars being a luxury we can't afford but rather a necessity we can't afford to be without. The human race *needs* an expedition to Mars.

Turn the clock back forty years to the Apollo era or one hundred years back to the age of polar exploration and you will see a phase in human history when nations were brought together by challenges, and when people knew what it was to feel a part of something and to have heroes. In the twenty-first century, in the absence of the challenges faced by polar explorers or those confronted by the Apollo astronauts, humanity needs to regain its self-worth as a species and redefine itself by once again rising to a challenge. The challenge should be a manned mission to Mars.

Surviving such a trip will not be easy, but neither was reaching the South Pole almost a hundred years ago. Landing humans on Mars will be a tough and arduous mission but no more so than the successful expeditions of Roald Amundsen and Fridtjof Nansen. Just as Amundsen and Nansen were tested, so too will a mission to Mars test the bravery, endurance and ability to cooperate of those fortunate enough to be chosen for such an expedition.

THE ROLE OF NASA

Despite the daring legacy of the Apollo missions, NASA never was an agency known for throwing caution to the wind. In fact, it took President Kennedy to direct NASA towards the Moon and even then, many of the space agency's scientists and officials were opposed to the idea of landing humans on the Moon, arguing the technology was unproven and worrying whether it would be safe to land on the Moon's surface.

Instead, the naysayers argued, it would be more prudent to wait until the technology had been proven rather than risk an embarrassing failure.

The cautious approach of NASA has been evident over the last three decades as astronauts have been confined to LEO operations using the less-than-ideal Space Shuttle to visit the problem-ridden International Space Station (ISS). Today, NASA's deeply-ingrained safety culture is even more entrenched in the wake of the 2003 Space Shuttle Columbia tragedy, a reason the agency wants to wait until 2031 before attempting to land humans on Mars. NASA's biggest argument against sending people to Mars *now* is that not enough is known about how to prevent bone demineralization, protect astronauts from radiation exposure during such long missions or how to ensure their safety during the dangerous entry, descent and landing (EDL) phase.

Of course it is hard to stand up against safety in a world where cars are equipped with multiple airbags and antilock brakes and where people hold McDonalds responsible for scalding them with coffee that is too hot! Unfortunately, such a safety culture is the exact opposite to many people's ideas of what explorers should do. Historically explorers risk their lives and sometimes they die. When the Space Shuttle Columbia disintegrated on re-entry on February 1, 2003, there were some people who argued the manned space program was just too risky and should be shut down. The reality is that risk-free exploration does not and cannot exist. Barring a Presidential edict however, it is difficult to see how NASA would accelerate its Mars program. That said, NASA is probably the one space agency that can succeed in attaining the goal of landing humans on Mars.

One reason is that NASA is a national institution that enjoys an extraordinarily positive approval rating. The American public love NASA, pure and simple, and even if the average American may not be sure exactly what NASA does, such strong brand loyalty is an asset for the agency's future because such support has an important bearing upon funding and space policy-making. While there may never be another "Man, Moon, decade" speech, NASA will succeed because they have succeeded in so many other space endeavors. For all its political flaws, the construction of the ISS was mainly a NASA project and its near-completion represents the greatest engineering project in the history of mankind. Furthermore, the construction of the ISS was a multi-decade achievement akin to what will be required for a manned mission to Mars.

THE INEVITABILITY OF HUMANS ON MARS

Inevitably, humans will venture to Mars and other planets within the Solar System. These missions will occur because humans are driven by a voracious yearning for exploration and because the technical resources to accomplish these objectives are available. The question is not whether humans will embark upon these missions but *when* and *how* these expeditions will take place. Just as polar expeditions led by Ernest Shackleton and Fridtjof Nansen were unique, so too will the two-to-three-year missions contemplated by NASA and the European Space Agency (ESA). How will these missions be designed? How will spacecraft survive the harrowing the Martian entry, descent and landing sequence? How can potentially serious and perhaps mission-threatening behavioral issues be avoided? How can the insidious effects of bone demineralization and radiation sickness be avoided? The answers to these questions unfold in the pages that follow.

REFERENCES

1. Next Generation Exploration Conference. Mars Science and Exploration Working Group. NASA Ames Research Center, August 16–18, 2006.

2

Interplanetary plans

"International cooperation expands the breadth of what any one nation can do on its own, reduces risks and increases the potential for success of robotic or human space exploration initiatives. It is important to establish and sustain practical mechanisms to support exploration if humanity is to succeed in implementing long-term space exploration on a global scale."

Statement made by the Framework for Coordination
of the Global Exploration Strategy

NASA and the European Space Agency (ESA) both have a plan to send humans to Mars. NASA's plan is known as *Constellation* and ESA's equivalent is *Aurora*. Russia, though its plans are less well defined than those published by NASA and ESA, has announced its intentions of sending cosmonauts to the Red Planet by 2035, a mission it may undertake in collaboration with China. Although it may be possible for either NASA, ESA or a Sino-Russian venture to embark upon a unilateral Mars mission, in reality international collaboration will be required to make such exploration sustainable. This fact is reflected in the Global Exploration Strategy (GES), a template that may result in one or more nations joining forces in their efforts to land astronauts on the surface of Mars. Meanwhile, NASA and ESA, and to a lesser extent, Russia and China, continue to develop the technologies required to realize a manned mission to Mars in the 2030 to 2040 timeframe. This chapter discusses the plans of the major international space agencies to develop these technologies and assesses the political posture for embarking upon a manned Mars mission.

EUROPEAN SPACE AGENCY

Endorsed by the European Union Council of Research and the ESA Council in 2001, *Aurora* is a part of Europe's strategy for space exploration, calling for Europe to explore the Solar System, stimulate new technology and inspire young Europeans to take a greater interest in science and technology. Also within the Aurora program is a plan to embark upon robotic and human exploration, a strategy that includes the clearly-defined objective of landing humans on Mars and returning them safely.

Since Aurora is an ESA program, it is envisaged international cooperation will be involved in many of the planned missions, especially one as complex as sending humans to the Red Planet. To achieve such a complex goal, Aurora's exploration strategy, in common with NASA's Constellation program, is based on an incremental approach focused on increasing mission complexity over time. Implementing this approach will begin with remote sensing of the Martian environment followed by robotic exploration and surface analysis missions. A Mars Sample Return Mission (MSRM) will then be followed by a mission establishing a robotic outpost, after which the ultimate goal of sending humans to Mars will be realized.

Although the immediate focus is on robotic missions, ESA is already conducting preliminary planning for human exploration, utilizing a mediated and integrated design effort involving members of industry and national space agencies. For example, aerospace companies Alenia Spazio, Astrium and Alcatel are each conducting a parametric analysis of a human mission to Mars (Table 2.1), an approach ESA hopes will identify comparisons in mass, crew number (3, 6 or 12 crewmembers) and duration of stay (30, 100 or 600 days) for such an expedition before commencing the initial mission architecture studies. Such studies will also enable ESA to define other mission constraints such as cost and mission objectives and assist the agency in determining how far its present day assets in human spaceflight need to be advanced before it can decide whether to embark upon a manned Mars mission.

Table 2.1 Manned Mars mission parametric studies

Launchers	Martian logistic infrastructure	Strategy for Mars arrival
Assembly in Orbit	Resources required by crew	Entry, descent and landing
Communications	Quarantine & Planetary protection	Ascent from Mars surface
Navigation	Crew environment (dust, radiation)	Power for Mars base
Propulsion	Earth-Mars-Earth Trajectories	Earth re-entry

Aurora missions

The primary objectives of the Aurora missions include gathering information about the Martian environment, the risks associated with it and with the interplanetary journey to and from the planet, in addition to developing and validating key enabling exploration technologies. To realize these objectives, an exploration roadmap (Table 2.2) of enabling technologies has been developed with a series of key milestones represented by specific types of missions (Table 2.3).

Table 2.2 European Space Agency exploration roadmap

Phase	Timeframe	Description
1	**Up to 2020**	Advancement of human operations in LEO based on utilization of the ISS. Development of new generation of crew space transportation systems, designed for LEO and low lunar orbit (LLO). Early robotic preparatory missions towards the Moon and Mars. Technology demonstrators of planetary descent and landing.
2	**Up to mid 2020s**	Extended human operations in LEO. Orbital infrastructures beyond LEO constructed as an element of transportation architecture. First Mars Sample Return Mission implemented. First human missions to the Moon.
3	**Late 2020s to early 2030s**	Extended lunar surface installations for fixed and mobile habitation and research introduced. Activities towards preparation of an international mission to Mars. Commercial services integrated part of space activities. Sustained, long-term presence on Moon.
4	**Mid to late 2030s**	Implementation of first human mission to Mars. Commercial enterprises operating in LEO. Continuation of lunar activities.

Table 2.3 Aurora exploration roadmap - mission by mission

Mission	Timeframe	Primary Objectives
Exomars	**2013**	Demonstrate entry, descent and landing of large payload on the surface of Mars. Technology demonstration of surface mobility via a rover. Search for signs of past and present life. Automatic sample preparation and distribution for analysis by scientific instruments.
Robotic Missions	**2015–2016**	**Exploration Lunar Orbiter** Provide information on the Moon environment for later missions. Define future safe paths for rovers. Provide data for manned landing sites.
	2017 – 2018	**Demo Lunar Lander** Technology demonstrator for life support and materials exposure. Technology demonstrator for robotic systems, communications and navigation. Technology demonstrator for soft precision landing.

Table 2.3 *cont.*

Mission	Timeframe	Primary Objectives
Crew Orbital Missions	**Up to 2020**	Demonstrate assembly in orbit for planetary missions. Demonstration of life support aspects of Moon mission in LEO. Demonstrate Mars docking system in LEO. Demonstrate re-entry vehicle from LEO.
Crew Orbital Missions. CTV	**Up to mid-2020s**	**Secure crew access to existing and future orbital research** infrastructures in LEO. Enable participation in human exploration.
Mars Sample Return	**Up to mid-2020s**	Search for signs of life. Perform geological analyses on samples of Martian soil. Entry, descent and landing system validation and Mars ascent vehicle validation. Operational aspects of round trip to Mars.
Cargo Lander	**Up to mid-2020s**	Science and technology demonstration. Delivery of logistics to lunar base. Provision of consumables for extended surface exploration. Delivery of surface assets to support lunar outpost.
Mars Robotic Lander	**Late 2020s to early 2030s**	Soft landing payload delivery on Mars surface. Topographical mapping. High resolution imagery. Martian weather monitoring and weather prediction.
Human Moon Mission	**Late 2020s to early 2030s**	Demonstration of life support system. Return crew from lunar surface to orbital infrastructure. Demonstration of re-entry to Earth. Perform trans-lunar maneuvers.
Cargo Element of First Human Mission	**Mid to late 2030s**	Technology demonstration of inflatable module. Provide a storm shelter for solar particle events. Sustains crew on surface and provides an interface to perform surface activities.
First Human Mission to Mars	**Mid to late 2030s**	Land a crew on Mars and return them safely, ensuring planetary protection for Earth and Mars. Demonstrate human capabilities needed to support human presence on Mars. Perform exploration and expand scientific knowledge taking maximum advantage of human presence. Sustain crew of four for up to 2.5 years in deep space.

Figure 2.1 An artist's rendering of the European Space Agency's Mars Sample Return ascent module lifting off from the Martian surface. The ascent module will launch into orbit, rendezvous with the Earth re-entry vehicle and return to Earth on a ballistic trajectory carrying precious samples from Mars. (European Space Agency.)

ExoMars

ExoMars (Figure 2.1), due to be launched in 2013, will demonstrate a number of candidate technologies that may be incorporated into a human Mars mission. For example, the ExoMars descent module will use an inflatable braking system or parachute system to decelerate during the entry, descent and landing (EDL) phase. Once on the surface, the lander will deploy a high-mobility rover carrying a comprehensive suite of scientific instruments whose objectives will be to search for traces of past and present life and identify surface hazards for human missions. The descent module will also carry an exobiology payload, including a lightweight drilling system and a sampling and handling device capable of performing in-situ soil analysis. During the mission, rendezvous and docking maneuvers will be tested by means of a rendezvous experiment (RVE) that will validate rendezvous and docking technology (RDT), a critical element in a future manned mission.

Mars Sample Return Mission
The MSRM is scheduled for launch in Phase 2 of the Aurora timeframe. The MSRM will comprise no less than five spacecraft, including an Earth-Mars transfer stage, a Mars orbiter, a descent module, an ascent module and an Earth re-entry vehicle. In terms of mission complexity, the MSRM will be almost as challenging as a manned mission. In common with the ExoMars mission and with the future manned mission, the MSRM will utilize an inflatable braking device to enter into Mars orbit, whereupon the descent module will be released and descend to the Martian surface. Once samples of soil have been collected, they will be loaded onto the Mars ascent vehicle and launched into orbit where the ascent vehicle will rendezvous and dock with the Earth re-entry vehicle, which will then return to Earth where the samples will be recovered.

MSRM will be a pioneering mission designed to validate a number of new technologies that must function perfectly if a manned mission is to be realized. These include the type of landing system to be used on Mars, the design of the Mars ascent vehicle, the rendezvous and docking operation in Mars orbit and the Earth re-entry vehicle.

The MSRM will lay the foundation for the Phase 3 suite of missions intended to prepare for an international mission to Mars by first demonstrating a sustained long-term presence on the Moon. Once this has been achieved, Phase 4 will see the implementation of the first human mission to Mars, but whether this mission is a European venture or in partnership with Russia and/or the United States remains to be seen.

European politics

> "I am convinced that an exploration program can only be global, without
> exclusivity or appropriation by one nation or another."
> *French President, Nicolas Sarkozy, during a visit to ESA's*
> *launch site in Kourou, French Guyana.*

Officially, European space is in great shape, but a space summit at Kourou, French Guyana in July 2008, suggested stronger political guidance is required to define new European space goals in the long-term since support for ESA's objectives is far from unanimous. One problem identified by those attending the summit was the loss of European focus on cooperative programs, a factor crucial to achieving a large-scale program of landing humans on Mars. Although French President, Nicolas Sarkozy, attempted to prod member states into drawing up a road map and timetable geared to eventually landing astronauts on Mars, the reality is that under a new European Union (EU) treaty, it is now the European Parliament that has the right to initiate legislation on space matters. This means ESA member states must succeed in making space a high-level political issue, something it has never enjoyed. To that end, a rising chorus of voices is insisting Europe must develop its own independent manned space transportation-system capability. After abandoning the Hermes spaceplane in 1993, ESA has had to rely on the Space Shuttle and the Russian Soyuz to ferry its astronauts to the International Space

Station (ISS) and, if ESA is to progress to landing humans on Mars, it will need to have its own independent manned vehicle.

RUSSIA AND CHINA

Despite the various hardships the Russian space program has endured since the dissolution of the Soviet Union, there are still many who dream of going to Mars. The S.P. Korolev Rocket and Space Corporation, also known as Energiya, even has a website devoted to a proposed manned Mars mission, despite the reality that the era of independent Russian interplanetary exploration is over, at least for the time being.

Russia has wanted to go to Mars since the late 1960s, once it became clear the race to the Moon was lost. Since losing out to the Americans, the Russians switched their focus to long-duration space flight, continuing to man the Mir space station until it was de-orbited in 2001. During the Mir era, the Russians gained invaluable experience in supporting extended-duration missions, some of which lasted more than a year. In fact, the record for the longest spaceflight ever is held by Russian cosmonaut, Valery Polyakov (Figure 2.2), who spent four hundred and thirty seven days in orbit between January 8, 1994, and March 22, 1995.

Figure 2.2 The Russian cosmonaut, Dr Valery Polyakov, who holds the record for the longest single space flight (437 days). (www.spacefacts.de)

Since the de-orbiting of Mir, manned Mars missions have been the subject of talk and little else, although Nikolai Sevastyanov, ex-president of Energiya, optimistically believes a manned Mars project could be achieved after 2025. Consisting of three stages, the Russian route to Mars would begin with a trial expedition around the Moon, followed by a non-landing manned expedition to Mars and, finally, a manned Mars landing. Unfortunately, despite the regular announcements in the press of impending Mars missions, the reality is the Russian space agency probably couldn't afford such an endeavor.

Although Russians hope to set foot on Mars, it is likely this goal will only be achieved with the cooperation of other space agencies. Such cooperation may be with ESA or NASA but it may also be with either India or China, each of which has been involved in discussions with the Russians. In September, 2007, Russia and India held discussions on the possibility of cooperation on missions to the Moon and to Mars and, earlier in the year, the Chinese National Space Administration (CNSA) and the Russian Federal Space Agency agreed to launch a mission to Mars as early as October 2009, albeit a robotic one. The Sino-Russian mission intends to launch a Chinese satellite to the Martian moon Phobos, where soil samples will be collected and returned to Earth. Whether the mission is realized is perhaps less important than the agreement indicating the two sides have taken a key step forward to working together on a space program.

The partnership of China and Russia came as no surprise to veteran Sino-Russian observers. The combination of Russian technology and its long-duration space exploration experience when merged with the Chinese economy clearly make such a pairing a win-win situation. In fact, China's space program can be traced back to the mid-1950s, when it was started with Soviet assistance during a period of strong ties between the two Communist bloc giants. Now, it seems those ties have been renewed, as evidenced by the aforementioned agreement and Russian assistance in the advanced training of China's taikonauts. Already, China has undertaken its third manned spaceflight (Figure 2.3), has completed development and testing on a new extravehicular (EVA) spacesuit and has published plans for a twenty-tonne human-tended space station. Supporting these projects are more than two hundred thousand engineers involved in aerospace research and development work in areas such as propulsion, robotics, space nuclear power and a host of other technologies required to operate in space, be it in LEO or on the surface of Mars.

How far this strategic partnership and cooperation will go in terms of realizing a Sino-Russian manned mission to Mars is uncertain, but Russia's unique scientific capability and experience in long-duration spaceflight together with China's funding has the potential to mark a revolution in manned spaceflight. If that happens, it is possible the centre of gravity for space exploration and a future manned mission to Mars may begin to move from the Atlantic to the Pacific.

Figure 2.3 Launch of Shenzhou 6, the second Chinese manned space mission. The third manned space mission, Shenzhou 7, launched in September 2008, demonstrated China's extravehicular activity capability when it became the third nation to conduct a spacewalk. (NASA.)

UNITED STATES

> "We do not know where this journey will end, yet we know this – human beings are headed into the cosmos. Mankind is drawn to the heavens for the same reason we were once drawn into unknown lands and across the open sea. We choose to explore space because doing so improves our lives and lifts our national spirit."
>
> *President George W. Bush, unveiling what he billed as a new course for the nation's space program, in a speech at NASA Headquarters, January, 2004.*

In January 2004, President George W. Bush launched a program to return humans to the Moon by 2020 and then to Mars, by a date yet to be determined, although February 2031 has often been mentioned. Although that date is a long way in the future, NASA is already planning to start development work on the lander vehicle and interplanetary transfer stage that will be used for a manned mission. In the fourth quarter of 2009, as part of NASA's $125-million Propulsion and Cryogenics Advanced Development (PCAD) project, almost six million dollars will be spent on Mars propulsion technologies for the aforementioned vehicles. The program responsible for developing the hardware and related exploration architecture systems that will eventually transport humans to Mars is *Constellation*. The capabilities developed under Constellation are similar to ESA's Aurora program and include robotic precursor systems, crew and cargo transportation, surface systems, in-space systems and ground systems. These capabilities will be developed within five defined program *spirals* (Table 2.4), ultimately leading to humans landing on Mars as the goal of Spiral 5.

The new vision

When President Bush announced NASA would send astronauts to the Moon by 2020 as practice for a manned landing on Mars, the idea was dismissed by many as either election-year grandstanding or an attempt to cast the President as a figure of history. Others wrote off the President's vision as an unserious proposal with little chance of succeeding.

However, for NASA, still reeling from the tragedy of the Space Shuttle Columbia disaster less than twelve months before, the new Vision of Space Exploration (VSE) offered the beleaguered agency a chance to recapture its faded glory. Even though the mission to land humans on Mars is expected to take two or more decades, the vision is proceeding, a new family of launch vehicles is being developed and the Crew Exploration Vehicle (CEV), named Orion (Figure 2.4), is being constructed.

To realize the goal of landing humans on Mars will require engineering on a scale of difficulty greater than anything ever accomplished but if one country is capable of such extraordinary feats, it is the United States. The task of returning humans to the Moon and ultimately traveling to Mars is now the defining focus of NASA, a goal that has not only refocused the "can do" spirit of the agency but is one the public can relate to. The question now is whether the agency can achieve the goal within a

Table 2.4 Constellation spirals

Spiral	Mission	Timeframe	Primary Objectives
1	4-6 crew to Low Earth Orbit	2014	Demonstrate operational qualification of crew exploration vehicle. Demonstration of Earth entry and recovery. Demonstration of aborts. Demonstration of automated control.
2	4-6 crew to lunar surface for extended duration	2015-2020	Demonstration of Earth to Moon cruise capability. Demonstrate low lunar orbit operations. Demonstrate untended lunar orbit operations Place humans at lunar base camp.
3	4-6 crew to lunar surface for long-duration stay	2020-TBD	Demonstrate lunar surface operations capability. Demonstrate lunar habitat operations. Increase system technical performance. Provide increasing levels of operational autonomy capabilities required for Mars mission.
4	Crew to Mars vicinity	2025 +	Demonstrate Earth to Mars cruise capability. Demonstrate Mars vicinity operations. Demonstrate Mars to Earth cruise capability.
5	Crew to Mars surface	2030 +	Land a crew on Mars and return them safely. Demonstrate human capabilities needed to support human presence on Mars.

reasonably engaging timeframe and within a budget that doesn't leave taxpayers complaining, thereby forcing the government to abandon the idea altogether.

American politics
The ISS had its genesis in a proposal made by President Reagan in his 1984 State of the Union address to construct a space station in low Earth orbit (LEO). The ISS was originally scheduled to have been completed in 1994 but in 2008, it is still several Space Shuttle missions away from being finished. The ISS also missed the mark in terms of budget since it was supposed to have cost only eight billion dollars but the 2008 estimate is closer to one hundred billion. It is hardly surprising, therefore that some point to the 'over budget and behind schedule' ISS as a model for predicting how a manned Mars mission may fare. However, it is unlikely a manned Mars mission would follow the model of the ISS, which, due to changing political goals, suffered several modifications that often did more harm than good. Furthermore, the decision to allow Russian participation resulted in unfortunate fiscal and scheduling problems that hampered construction. While it may be unlikely, given the recent alliance of Russia with China, the United States will collaborate with their old space station partner to realize a manned mission to Mars, it is possible NASA may

Figure 2.4 Orion is NASA's replacement for the Space Shuttle and is due to enter service in 2015. It will also be the vehicle carrying Mars crews to low Earth orbit. (NASA.) See Plate 2 in color section.

invite ESA to participate in such a venture. However, if such a coalition is not realized, NASA remains the only space agency capable of going it alone, budget challenges and a change of administration notwithstanding.

GLOBAL EXPLORATION STRATEGY

"Scientific discovery and technological development drives human advancement. Mars provides a unique and vital destination for such research. In addition, the exploration of Mars can provide significant social benefits back on Earth. International cooperation will not only be essential to the success of a human presence on Mars, but development of such international collaboration processes would jumpstart collaboration on other global issues. The eventual commercialization of space holds tremendous opportunities for economic growth. Furthermore, there is an undeniable basic human need to explore and define our place in the universe. The overarching theme that ties together all of these reasons for exploration is inspiring and uniting the global community. Continuously inspiring the public, the scientific community, and the community of Earth is required to achieve the goals of Mars science and exploration."

Declaration of the Mars Science and Exploration Working Group.

Although NASA may place its astronauts on the surface of Mars before Russia, China or Europe, sustainable space exploration is a challenge no one nation can pursue on its own. This is why fourteen space agencies (Table 2.5) have developed the GES: The Framework for Coordination, representing a vision for robotic and human space exploration.

Although the Framework does not focus on a single global space mission, it recommends a non-binding forum through which space agencies can collaborate to strengthen individual projects. Obviously a prime target of the Framework would be a manned mission to Mars, since the complexity of such an expedition lends itself to an international coordinated strategy. Ideally, such a strategy would create a common language of exploration building blocks, such as spacecraft interoperability and common life support systems. The partnership approach has already been demonstrated in the undertaking of the largest space project ever, namely the construction of the ISS. In constructing the ISS, the United States, Canada, Europe, Japan and Russia achieved together what no one nation could have accomplished alone and, in the process, forged strong political ties that may prove valuable when it comes to embarking upon a manned mission to the Red Planet.

Table 2.5 Members of the Global Exploration Strategy: The Framework for Coordination

Agency	Country	Agency	Country
Agenzia Spaziale Italiana (**ASI**)	Italy	European Space Agency (**ESA**)	Europe
British National Space Centre (**BNSC**)	United Kingdom	Indian Space Research Organization (**ISRO**)	India
Centre National d'Etudes Spatiales (**CNES**)	France	Japan Aerospace Exploration Agency (**JAXA**)	Japan
China National Space Administration (**CNSA**)	China	Korea Aerospace Research Institute (**KARI**)	Korea
Canadian Space Agency (**CSA**)	Canada	National Aeronautics and Space Administration (**NASA**)	USA
Commonwealth, Scientific, Industrial Research Organization (**CSIRO**)	Australia	National Space Agency of Ukraine (**NSAU**)	Ukraine
Deutsche Zentrum für Luft und Raumfahrt (**DLR**)	Germany	Roscosmos	Russia

3

Mission architectures

Perhaps the most important decision of a manned Mars expedition will be choosing a mission design. A mission design, or *mission architecture*, consists of several complex and inter-related variables such as propulsive maneuvers, crew support, aero assist requirements, habitation, and atmospheric entry system requirements. Mission architecture also describes a combination of modes specific to the mission, the assigning of flight elements such as the launch vehicle and, in the case of establishing a Martian outpost, a description of the activities performed on the surface. Before describing some of the many mission architecture suggested for a manned Mars mission, it is useful to understand some of the elements common to each.

INTERPLANETARY TRAJECTORIES

To understand the subtleties of interplanetary travel, it is necessary to have a basic understanding of planetary motion and orbital mechanics. On a voyage to Mars, the vehicle and the celestial bodies are in constant motion and nothing travels in a straight line which means the timing of any interplanetary mission design assumes critical importance, as does the calculation of the propellant necessary to execute the required trajectory. In addition to calculating propellant and executing trajectories, mission designers must also consider the celestial bodies exerting a gravitational influence upon the spacecraft. For a trip to Mars, these include the Earth, Mars and the Sun although, in actuality, the spacecraft will also encounter gravitational effects due to Venus, solar pressure and other perturbations [13].

Basic orbital mechanics
The Earth and Mars move in elliptical orbits (which are very nearly circular) and in almost the same plane. A spacecraft traveling to Mars must follow an elliptical path from Earth orbit to the orbit of Mars. To begin this journey, a spacecraft is placed in low Earth orbit (LEO), from where it fires its thrusters to change its velocity, or delta V [13]. The change in velocity launches the craft into a hyperbolic escape from LEO at a velocity of at least 4.8 kilometers per second, sending the spacecraft on an

elliptical path until it intersects an imaginary point in space where the gravity of Mars begins to dominate. Along the way, mid-course corrections may be applied to ensure the spacecraft enters the orbit of Mars, rather than crashing directly into the planet.

Trajectory variables

In deciding on the trajectory mission architects must consider several variables, the first of which is the amount of energy required to accomplish the trip. To calculate the amount of fuel, it is necessary to sum the changes in velocity required to leave LEO, enter Mars orbit, leave Mars orbit and enter LEO. This value, or *total delta V*, is used to calculate the total amount of fuel required to complete the return trip. Obviously, since fuel adds to the cost and weight of the mission, decreasing the fuel is a priority for mission designers but the desire to save fuel must also be balanced against the safety of the crew. As we shall see in Chapter 8, an interplanetary flight is fraught with health hazards posed by radiation and bone demineralization, so the crew must spend as little time in zero G as possible but as trip duration decreases, the power required to complete the trip increases due to increased fuel expenditure. Since fuel expenditures are directly proportional to the payload mass it is not feasible to send all the scientific equipment, human support systems and the crew on the same flight. Therefore, most mission architectures comprise a two or three-flight plan in which communications and scientific equipment are sent first on a slower, lower power trajectory and a human crew is launched on a second, faster, higher power trajectory.

Trajectory options

At its closest approach to Mars, the distance from the Earth is fifty-six million kilometers whereas, at its farthest, this distance increases to about four hundred million kilometers. When deciding how to cover this distance, the mission planner has a choice between a Hohmann Transfer Trajectory, an Opposition Trajectory or a Conjunction Trajectory.

Hohmann transfer trajectory

The cheapest and slowest travel route to Mars utilizes the *Hohmann trajectory* (Figure 3.1), named after the German engineer and mathematician, Walter Hohmann, in 1925. Utilizing this trajectory, the spacecraft departs LEO in the same direction as the Earth is traveling and arrives at Mars nearly nine months later, on the opposite side of the Sun from where the spacecraft left Earth. Because the thrust of a spacecraft utilizing the Hohmann trajectory uses very little fuel, the trajectory is often referred to as a *minimum energy transfer*. Although such a long travel time is not suitable for human missions, the Hohmann option, which requires the spacecraft to travel the greatest distance to Mars, is a useful way of delivering cargo to the planet.

Opposition trajectory

Another trajectory available is the planetary flyby, or opposition trajectory mission, a transfer in which the spacecraft bound for Mars first flies past Venus, in a 'gravity

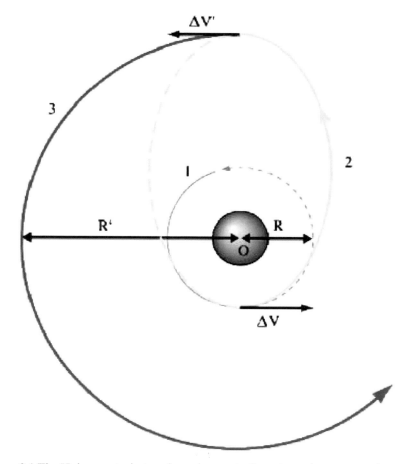

Figure 3.1 The Hohmann trajectory is analogous to throwing a dart at a moving target, where it is necessary to lead the aim point by just the right amount to hit the target. In this diagram, the Hohmann transfer orbit is one half of an elliptic orbit linking the low Earth orbit (numbered 1 on the diagram) and Mars orbit (numbered 3 on the diagram). The transfer (numbered 2 on the diagram) occurs as the result of the spacecraft performing the trans-Mars insertion (TMI) burn accelerating the vehicle so it follows the elliptical orbit. Once it reaches Mars the vehicle fires its engines again to attain orbital velocity around Mars. (Wikimedia Commons, based on image by Hubert Bartkowiak.)

assist' maneuver, 'stealing' energy from the Venusian gravitational field, before continuing on a (modified) trajectory taking it on to Mars [13]. Opposition trajectories provide significantly shorter surface times in the order of between thirty and ninety days, in-space durations as long as one and a half years and relatively large propulsive requirements. This type of mission is also the most complex due to the Venus swing-by. Unfortunately, the flyby option is not a viable alternative for human missions because the transfer time is increased significantly and the trajectory brings the spacecraft too close to the Sun, which means the crew is exposed to a high level of solar radiation.

Conjunction trajectory
Finally, there is the conjunction trajectory, characterized by long surface times in the order of four hundred to six hundred days, short in-space durations of approximately one year, and relatively minor propulsive requirements. A conjunction class mission also offers the crew a 'free-return' trajectory, which is one of the abort options discussed in Chapter 4. A free-return trajectory, which may be used in the event of a contingency event mandating a return to Earth, simply requires the vehicle to enter Mars orbit and use the gravity of the planet to return without expending any extra fuel, hence the term 'free-return'. Since the conjunction-class option requires less shielding and propulsive requirements and is less risky due to the reduced time in zero gravity, it is the mission of choice of most mission designers.

Conjunction trajectory options
Assuming mission planners choose the conjunction trajectory option, the next step is to calculate the Earth departure delta-V, which in turn will determine the time it will take for the spacecraft to travel to Mars. Simply providing as much thrust as possible won't work because once the spacecraft arrives at Mars it needs to brake and if it is traveling too fast it will not be able to capture into Mars orbit. A reasonable delta-V of 3.98 kilometers per second will enable a transfer of 200 days [13], whereas decreasing the transit time to 180 days requires a delta-V of 4.08 kilometers per second, a velocity still within the current capabilities of braking into the Martian atmosphere. The choice of which departure delta-V not only impacts how long the crew can stay on the surface of Mars before returning to Earth but also determines which abort options are available to the crew. Generally, faster Earth-Mars transit times translate into longer Mars stay times, while high Earth departure delta-Vs are characterized by longer free-return trajectories, and the shorter free-return trajectories suffer from high Mars entry velocities.

Braking into orbit
Another important component of any mission architecture is the method for braking a spacecraft into the orbit of Mars, an event achieved utilizing an aero assist trajectory (AAT). Braking into orbit forms a part of the entry, descent and landing (EDL) phase of the mission, which is described in detail in Chapters 4 and 9. However, for the purposes of this chapter, only a basic overview of the methods is required.

Aero assist trajectory
An AAT is one in which the spacecraft uses the Martian atmosphere to non-propulsively alter its velocity, thereby saving propellant. This results in a significant reduction not only in the initial mass of the Mars crew vehicle but also in launch cost and architecture complexity. The aero assist maneuvers required by a Mars mission

are performed on Mars arrival and on Earth return and are achieved by placing the vehicle on a trajectory that dissipates vehicle energy through heat transfer.

A high-speed steep entry into an atmosphere will result in the largest deceleration but also the highest heat rates, whereas a high-speed shallow entry will result in less deceleration and a lower heat rate that lasts longer. In choosing the optimum trajectory, mission planners must consider the aerodynamics of the vehicle and the ruggedness of its thermal protection system.

In summary

This section has described the elements common to the mission architectures discussed in the remainder of this chapter, the intent of which is to present mission plans utilizing a variety of contrasting propulsion methods, hardware and surface architectures. Some of the architectures discussed here, such as Das Marsprojekt, are presented to provide an historical perspective and represent the vision of just one person, whereas others, such as Project Troy and the reference mission of the MarsDrive Consortium, are more recent and represent the ideas of dozens of people. Finally, there are the architectures proposed by the space agencies, such as the European Space Agency's (ESA's) Human Mission to Mars and NASA's Constellation program, which symbolize the efforts of thousands of employees all trying to figure out how best to transport humans to Mars and return them safely.

DAS MARSPROJEKT

The first manned Mars mission architecture [12] was developed by legendary spacecraft designer, Wernher von Braun, in 1948 and subsequently published in German in 1953 as *Das Marsprojekt,* and then in English in 1962 as *The Mars Project.* Although the scale of von Braun's vision of reaching Mars may not be financially realizable in the near future, his calculations and technical estimates are as valid now as they were in 1948.

Mission architecture

Assembly of von Braun's Mars mission was envisaged to take eight months, requiring the services of forty-six reusable space shuttles. Collectively, the shuttles would perform nine hundred and fifty launches from a very eventful launch complex located at Johnston Island in the Pacific Ocean. Once assembled in orbit, von Braun's Martian armada (Table 3.1) would comprise seven passenger vehicles and three cargo ships, each weighing 372 tonnes. The passenger vehicles would be equipped with twenty-meter diameter habitation spheres to accommodate the ten crewmembers, while the cargo ships would each carry a two-hundred-tonne winged lander.

To place the armada on a trajectory to Mars, the fleet would fire their engines for sixty-six minutes which would consume more than three quarters of the ship's starting masses. The fleet would then follow a minimum-energy Hohmann trajectory, resulting in a transit time of two hundred and sixty days. On arrival at

Table 3.1 Das Marsprojekt mission architecture [12]

Mission Summary

Item	Value	Item	Value
Launch year	1965	Mars duration	443 days
ISRU	No ISRU	Inbound duration	260 days
Crew	70	Total Mission time	963 days
Outbound duration	260 days	Number of launches	950

Mission Hardware
Von Braun Passenger Ship

Item	Value	Item	Value
Crew Size	10	Mass	3720 tonnes
Length	41.0 m	Engine Propellant	Nitric acid/hydrazine[1]
Diameter	20.0 m	Engine Isp	297 seconds.

Von Braun Cargo Ship

Item	Value	Item	Value
Payload	395 tonnes	Mass	3720 tonnes
Length	64.0 m	Engine Propellant	Nitric acid/hydrazine[1]
Diameter	20.0 m	Engine Isp	297 seconds.

Von Braun Landing Ship

Item	Value	Item	Value
Crew size	17	Payload	12 tonnes
Length	22.0 m	Mass	200 tonnes
Diameter	4.7 m	Engine Propellant	Nitric acid/hydrazine[1]
Span[2]	153.0 m^2	Engine Isp	297 seconds.

1. Although nitric acid/hydrazine propellants are corrosive and toxic, they have the advantage of being storable without refrigeration.
2. The landing ship was an enormous glider. The wings would have had an area of 2810 square meters, enabling the ship to glide more than 12,000 kilometers or more than half way around Mars.

Mars, propulsive maneuvers would be performed to place the fleet into a circular equatorial orbit, 1,000 kilometers above the Martian surface. The fleet would then orbit Mars, permitting the crew to survey the Martian surface and select an appropriate landing site.

Once a landing site had been selected, one landing vehicle would undock, de-orbit and glide to a horizontal landing on one of the Martian polar ice caps. Since no robotic missions had made their way to Mars when von Braun devised his plan, it was assumed the polar ice caps were the only area that could provide a flat landing surface. Once on the surface, the crew would have no means of return to Mars orbit since, instead of an ascent stage, the landing vehicle would carry pressurized crawlers and a habitat. Once the base and landing site had been established, the two remaining vehicles would descend. These vehicles would be equipped with ascent stages, providing the means to return to Mars orbit. During surface operations, fifty crewmembers would remain on Mars while the remainder tended to the passenger ships in orbit.

After conducting four hundred days of surface operations, the two ascent vehicles would rendezvous with the seven passenger ships and the trans-Earth injection (TEI) maneuver would be performed, sending the crew on a two-hundred-and-sixty-day transit back to Earth. On arriving at Earth the passenger ships would brake into a circular orbit at an altitude of 1730 kilometers and the crew would be ferried back to Earth by the reusable space shuttles.

For many science fiction enthusiasts, von Braun's vision of traveling to the Red Planet still has appeal but to mission planners, who must work within tight budgets and exercise fiscal responsibility, Das Marsprojekt is just too cumbersome and unwieldy to ever be seriously considered. In an age that has witnessed space budgets regularly spiral out of control, what is needed is a lean, no frills mission that achieves the objective but doesn't bleed the taxpayer dry. Such an economic mission is Mars Direct.

MARS DIRECT

In sharp contrast to von Braun's 'Battlestar Galactica' approach, Robert Zubrin's Mars Direct design is one that uses only the bare minimum of equipment to safely reach Mars. Zubrin is the founder of Pioneer Astronautics, which conducts research and development on space exploration, but he is best known for his bargain basement plan to send astronauts to Mars, described in his book 'The Case for Mars: The Plan to Settle the Red Planet and Why We Must' [14]. The plan, known simply as Mars Direct, utilizes currently available technologies, small spacecraft, incorporates a high level of redundancy and, perhaps most importantly, is affordable.

Mars Direct architecture

Mars Direct begins with the launch of a Cargo Launch Vehicle (CLV) such as the Ares V (Figure 3.2), which deploys an upper stage containing a 40-tonne payload into a direct trans-Mars injection (TMI) trajectory. The payload consists of an empty methane/oxygen-driven two-stage ascent and Earth Return Vehicle (ERV), six tonnes of liquid hydrogen, a 100 kWe nuclear reactor mounted in the trunk of a methane/oxygen-driven truck, a set of compressors and an automated chemical processing unit [14].

Shortly after entering Martian orbit, the payload lands on Mars, the truck drives away from the lander and the reactor is deployed to provide power to the compressors and chemical processing unit. Hydrogen is then catalytically reacted with Martian carbon dioxide to produce methane and water. The methane is liquefied and stored via a process known as *methanation*, while the water is electrolyzed to produce oxygen and hydrogen via a process known as water

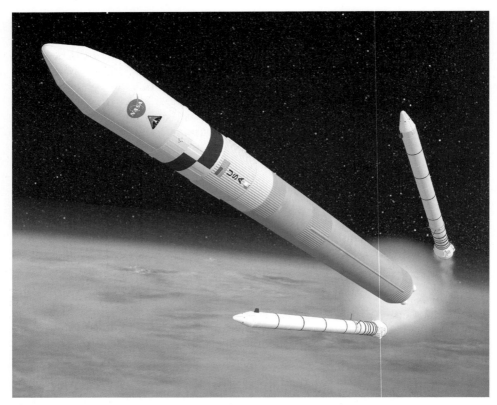

Figure 3.2 NASA's Ares V is a two-stage, vertically stacked launch vehicle capable of carrying 188 metric tonnes to low-Earth orbit. For the initial insertion into Earth orbit, the Ares V first stage utilizes two five and a half segment reusable solid rocket boosters derived from the Space Shuttle's solid rocket boosters. In the Mars Direct architecture, the Ares V would deploy a forty-tonne cargo payload into a direct trans-Mars trajectory. (NASA.) See Plate 3 in color section.

electrolysis. Utilizing these two processes, the six tonnes of hydrogen from Earth produce 24 tonnes of methane and 48 tonnes of oxygen. Furthermore, 36 tonnes of oxygen are produced via reduction of the Martian carbon dioxide, resulting in a total bipropellant production of 108 tonnes, of which 96 tonnes is used as fuel for the ERV and the remainder to fuel ground vehicles.

Once completion of propellant production has been confirmed, two more Ares Vs are launched. Each Ares V deploys a 40-tonne payload on a trans-Mars trajectory. One payload is an unmanned fuel-factory/ERV combination and the other is a habitation module containing four crewmembers, provisions sufficient for three years, a rover and an aerobraking/landing engine assembly. Shortly after arriving in Martian orbit, the manned payload lands in the vicinity of the first payload, where a fully fuelled ERV is available.

In the period between receiving confirmation of propellant production and the deployment of the second and third Ares Vs, the landing site is characterized by robotic exploration. Furthermore, the first payload assists the inbound crew by activating a transponder, thereby providing a targeting control system, enabling the pilot to fly a terminal descent to within a few meters of the intended location.

The inaugural Mars Direct crew remains on the Martian surface for 18 months. At the end of their surface mission, the crew returns to Earth directly, using the ERV. However, this is not the end of the mission, since the Mars Direct plan envisages missions occurring every two years in a strategy intended to leave behind a string of camps across the Martian surface.

Medical aspects

Artificial gravity

Among the several unique features of the Mars Direct architecture is the utilization of a tether system to create artificial gravity on the outbound phase of the mission. Following injection into the trans-Mars trajectory, the upper stage of the vehicle separates from the crew habitat and maneuvers to the other side of the habitat. As the upper stage moves away, it pulls the tether off the roof of the habitat and once the tether is extended to its full length of 1500m, the upper stage fires its reaction control thrusters and accelerates away at an angle. This maneuver gradually pulls the tether taut and begins to create artificial gravity (approximately 0.38 Earth-G, or one Mars-G) for both the habitat and the upper stage by rotating at one rotation per minute (RPM).

Why artificial gravity may not work

Although the utilization of a tether system represents an effective countermeasure against the deleterious effects of zero gravity, the design challenges of developing such as system remain formidable. Such a system would not only compromise communications between Earth and the spacecraft but also adversely affect the power gathered from the sun using solar arrays. Although maneuvers have been performed on spinning spacecraft such as the Pioneer Venus Orbiter, the challenges presented by incorporating such a system on a manned vehicle en-route suggest the tether option may not be a feature until later missions.

Surface architecture

One of the strengths of Mars Direct is its reliance on proven technology. For example, the in-situ production of methane and oxygen bipropellant is achieved using chemical processes that have been in large-scale use for more than a century. In the propellant production process, Martian carbon dioxide is compressed to about seven bar pressure and allowed to equilibrate to ambient Martian temperature, at which point the carbon dioxide condenses into a liquid state. From the liquid state, the carbon dioxide is vaporized and distilled in a process similar to the desalination method developed by Benjamin Franklin in the mid-1700s.

Once the carbon dioxide has been acquired, it is reacted with the hydrogen brought from Earth in the methanation (Sabatier) reaction developed in the 19th century. The methanation reaction has not only been used in industry for more than one hundred years but has also been the subject of research by NASA for possible use on the International Space Station (ISS).

In addition to being a proven process, the methanation reaction requires no energy because it is exothermic. As the reaction runs, the methane produced is liquefied and the water produced is condensed, transferred to a holding tank, before being pumped into an electrolysis cell and subjected to electrolysis. The electrolysis of the water produces oxygen, which is stored, and hydrogen, which is simply recycled back to the methanation reaction.

Mars Direct also utilizes the process of electrolysis in its use of solid polymer electrolyte electrolyzers to support the propellant-production requirements of the mission. In common with the robust history of Mars Direct's other energy production processes, solid polymer electrolyte electrolyzers have a reliable history of use onboard nuclear submarines and have acquired more than seven million cell-hours of experience.

Surface operations and exploration of the Martian surface are accomplished by a rover, powered by chemical combustion of propellant generated by the in-situ processes described previously.

Radiation and mission risk

The Mars Direct mission architecture uses a low energy conjunction class interplanetary transfer, a plan criticized by many as imposing excessive risk upon crewmembers due to the long exposure to radiation. However, Zubrin defends the choice by pointing out the deficiencies of the opposition class option. The large amounts of propellant required by a high energy opposition class mission, for example, require high energy aerocapture maneuvers, which in turn increase the risk of a spacecraft skipping out of the Martian atmosphere. Furthermore, Zubrin points out, the opposition class aerocapture maneuver would subject crewmembers to deceleration loads as high as eight Gs, which could prove disabling to deconditioned crews. Another shortcoming of the opposition class option, Zubrin argues, is its inefficiency, since this choice provides crews with only thirty days exploration time after an interplanetary round trip transit time of one and a half years.

On the subject of the risk of radiation exposure incurred by Mars Direct, Zubrin compares predicted transit and surface doses of minimum energy, Mars Direct and opposition class missions based on estimated worst-case doses (Table 3.2).

The pros and cons of Mars Direct

Zubrin's plan has a number of selling points. First, the robust mission architecture has a high level of redundancy built in, thanks to its cautious approach of sending the ERV ahead and telerobotically ensuring in-situ production of propellant has commenced. Secondly, the architecture, thanks largely to the in-situ propellant production, enables surface operations to commence as soon as the crew steps off the vehicle, thereby maximizing exploratory and science return. Thirdly, the architecture enables an incremental and continuous growth of a Martian outpost as a result of a new mission leaving Earth every two years.

Table 3.2 Radiation dose comparison [14]

	Minimum Energy	**Mars Direct**	Opposition
Transit[1] Doses (rem[2])			
GCR (Solar Minimum)	63.0	**45.0**	67.5
GCR (Solar Maximum)	5.9	**18.5**	27.8
Solar Flare[3] (Solar Minimum)	2.5	**1.8**	3.2
Solar Flare (Solar. Maximum)	12.7	**9.1**	16.0
Mars Doses			
GCR (Solar Minimum)	11.4	**14.3**	1.0
GCR (Solar Maximum)	5.5	**6.9**	0.5
Solar Flare (Solar Minimum)	1.1	**1.3**	0.1
Solar Flare (Solar Maximum)	5.4	**6.8**	0.5
Total Dose			
Solar Minimum	78.0	**62.4**	71.8
Solar Maximum	49.5	**41.3**	44.8

1. Based on average distance from Sun of 1.3 AU for conjunction and 1.2 for opposition.
2. Every 60 rem of radiation adds 1% of extra risk of a fatal cancer to a 35 year old female and every 80 rem adds 1% of extra risk of a fatal cancer to a 35 year old male.
3. Solar flare equal to average of three worst recorded events of Feb. 1956, Nov. 1960 and Aug. 1972. Assumes crew spends 25% of solar flare time unsheltered and 75% of time in shelter.

REFERENCE MISSION OF THE MARSDRIVE CONSORTIUM

Unsurprisingly, Zubrin's ground-breaking mission design not only inspired mission architects to develop more innovative plans but also encouraged others to refine the existing Mars Direct vision. Presented at the International Space Development Conference (ISDC) in 2006, the Reference Mission Design of the MarsDrive Consortium, also known as *Mars for Less*, is based on utilizing existing medium-lift launch vehicles, also referred to as Evolved Expendable Launch Vehicles (EELVs). Often described as a scaled-down version of Robert Zubrin's *Mars Direct*, the Mars for Less [1] architecture utilizes surface resources and long-duration stays to take full advantage of surface exploration time and exploration capabilities while on Mars.

Mission architecture

The core mission elements of Mars for Less include the unmanned Earth Return Vehicle (ERV) and the crewed Mars Transfer and Surface Vehicle (MTSV). The mission requires the ERV to depart first, followed by the MTSV two years later. Each vehicle comprises six components, delivered to LEO using EELVs. The first launch delivers a split-level habitat, to which a lander, aeroshield and other mission-specific modules are attached following the second launch (Table 3.3). The

Table 3.3 Earth return vehicle and Mars transfer and surface vehicle launches [1]

Earth Return Vehicle		Mars Transfer and Surface Vehicle	
Allocation: Launch #1	**Mass (t)**	**Allocation: Launch #1**	**Mass (t)**
Habitat Structure	3.0	Habitat Structure	3.0
Life Support System	1.0	Life Support System	1.0
Consumables	3.4	Consumables	9.0
Power	1.0	Power	1.0
Reaction Control System	0.5	Reaction Control System	0.5
RCS Propellant	3.3	RCS Propellant	3.0
Communications	0.1	Communications	0.2
Interior	0.5	Interior	1.0
EVA Suits (4)	0.4	EVA Suits (4)	0.4
Margin	1.5	Biomedical Package	0.2
Heat Shield	1.8	Crew (4)	0.4
Upper Propulsion Stage	1.5	Margin	2.5
Hydrogen Feedstock	6.3		
Chemical Plant	0.5	**Launch #1 Total**	**24.2**
Launch #1 Total	**24.8**	**Allocation Launch #2**	**Mass (t)**
		Cargo Module Structure	2.0
Allocation Launch #2	**Mass (t)**	Science Equipment & Lab	2.0
RTG	3.0	Open Rover	0.5
RTG Truck	0.4	Pressurized Rover	2.0
Lower Propulsion Stage	6.0	Margin	1.0
Aeroshield	6.2	Aeroshield	5.8
Lander (fully fuelled)	9.2	Lander (fully fuelled)	8.5
Launch #2 Total	**24.8**	**Launch #2 Total**	**21.8**
ERV Total	**49.6**	**MTSV Total**	**46.0**

remaining four launches deliver four propulsion stages which are mated aft of the vehicle and ignited in stages, resulting in a widening of the vehicle's orbit. With the final ignition of the last stage, sufficient energy is imparted to place the vehicle on a TMI trajectory (Figure 3.3).

The ERV proceeds to Mars on a near-minimum-energy transfer and aerobrakes into Mars orbit before landing with the aid of parachutes. Two years later, the MTSV follows a similar trajectory and arrives at Mars, following a four-to-six-month transit. Upon return, the crew boards the fully-fuelled ERV and launches from the Martian surface directly into an Earth-bound trajectory. On approach to Earth, the ERV uses aerobraking to enter LEO and the crew returns to Earth by means of a direct reentry, resulting in an ocean splashdown.

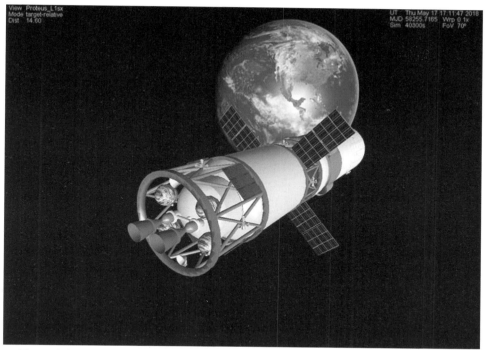

Figure 3.3 The Mars for Less stack departing for Mars following the trans-Mars injection burn. (Mark Paton, MarsDrive Consortium/Mars for Less.)

Mission hardware

A limiting design factor of the hardware used in Mars for Less is the requirement for the dimensions of the ERV and MTSV to fit into the payload shrouds of existing EELVs, achieved by limiting the exterior diameters of the vehicle to 4.5 meters.

The common habitat for the ERV and MTSV comprises a cylindrical, tapered structure dividing the vehicle into two levels. The habitable sections of the vehicle are connected by a half-cylinder running through the centre of the spacecraft. Level 1 contains a galley, sleeping quarters and a washroom and Level 2 consists of a laboratory, exercise compartment, additional sleeping quarters and command systems.

Since the vehicle is designed for a vertical flight and landing profile, there is only one primary load path, meaning structural reinforcement is limited to one axis only, permitting crucial savings in mass. More mass savings are realized by virtue of the ballistic atmospheric entry profile, an architecture design requiring less thermal protection.

Mission analysis

Since the ERV and MSTV would be assembled from twenty-five-tonne components, Mars for Less could take advantage of existing EELVs such as the Ariane V (Figure 3.4), the Delta IV-H, or one of the emerging launch vehicles such as SpaceX's Falcon 9-29.

Figure 3.4 The European Space Agency's Ariane V launch vehicle. (European Space Agency.)

While detractors of Mars for Less often cite on-orbit assembly as a disadvantage to any interplanetary architecture, the reality is that *orbital rendezvous, docking* and *assembly* are operations NASA, ESA and the Russian Space Agency have the *most* experience with. In fact, since the Mars for Less components would be compartmentalized, only minimal assembly would be required and any EVA work would be by several orders of magnitude less complex than that required to construct the ISS.

Another advantage of Mars for Less is the division of the spacecraft propulsion system, a design enabling the vehicle to incrementally discard parasitic mass. The propulsive performance of the vehicle is further enhanced by the use of staged TMI burns, conducted in the Earth's gravity well. This design feature of the architecture enables the vehicle to draw greater kinetic energy from each successive burn, saving energy through not accelerating a single upper stage through the whole TMI burn.

Perhaps one weakness of Mars for Less is the use of cryogenic bipropellants. Hydrogen and oxygen boil off when stored in space for long periods of time, but it is possible to reduce this boil-off to less than one percent per month by using special multi-layer insulation, consisting of aluminized Kapton and Dacron [14].

Another drawback to the multiple-launch architecture is the risk of launch delay and failures but this weakness is compensated for, to a degree, by the limited development costs, since there is no need for new launch technology and the resources required to implement the architecture already exist.

Ultimately, Mars for Less represents not only an alternative means to achieve a manned Mars mission, but also a realistic architecture that could conceivably be undertaken by private initiatives in the not too distant future.

PROJECT TROY

While Mars Direct and Mars for Less each utilize existing or soon to be realized hardware, other plans have adopted an approach developing an entirely new system. Such a plan is Project Troy [5].

Project Troy is a mission architecture suggested by Tony Martin, Richard Varvill and Alan Bond of Reaction Engines Limited, based in the United Kingdom. On reviewing previous mission architectures suggested by NASA and ESA, the Project Troy team considered the cost of lifting mission elements to orbit and the logistical challenges of the limited flight-rate of expendable vehicles, problems that may prove financially insurmountable. As an alternative, the Project Troy strategy for reaching the Red Planet utilizes the Skylon reusable Single Stage to Orbit (SSTO) spaceplane (Table 3.4, Figure 3.5), arguing if a mission to Mars is to be undertaken, there should be a practical spin-off. By incorporating the Skylon spaceplane into the mission architecture required to reach Mars, a reusable transportation system is created, thereby reducing the cost of spaceflight.

Figure 3.5 The Skylon Single-Stage-to-Orbit (SSTO) spaceplane in orbit. (Adrian Mann, Reaction Engines Limited.) See Plate 4 in color section.

Table 3.4 Skylon specifications [5]

Item	Feature
Length	82 m
Fuselage diameter	6.25 m
Wingspan	25 m
Unladen Mass	41,000 kg
Fuel Mass	220,000 kg
Maximum Payload Mass	12,000 kg
Powerplant	2 × SABRE combined ramjet, turbojet, rocket
Maximum Speed	Orbital
Service Ceiling	26,000 m (air-breathing). 200+ km (exoatmospheric).

Mission architecture

Project Troy comprises two mission phases, consisting of a total of six vehicles. In the automated *Precursor* phase, three vehicles transport equipment, surface habitats and power supplies to Mars two years prior to the second *Principal* manned phase, which ferries eighteen crewmembers in three spacecraft. This approach ensures a working surface base is operational in advance of the manned mission and also provides contingency options in the event of an aborted surface stay.

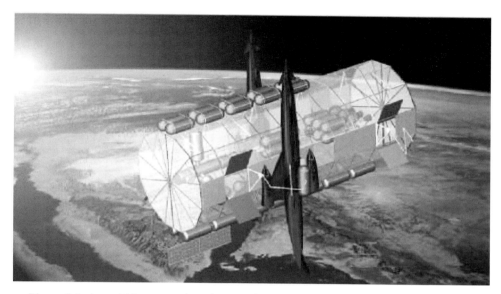

Figure 3.6 An orbiting operations base station in low Earth orbit will function as an integral part of the Project Troy mission architecture. (Adrian Mann, Reaction Engines Limited.)

The vehicles are assembled in low Earth orbit (LEO) at a space station facility, providing accommodation, assembly facilities and propellant storage (Figure 3.6). On completion, the vehicles depart LEO and enter a Hohmann transfer orbit (Table 3.5), permitting a fifteen-month surface stay on Mars and transit period of nine months.

Table 3.5 Project Troy Hohmann transfer parameters [5]

Phase I Precursor Mission	Departure date	Arrival date	Transit Time (d)
Outbound	05 Nov 2026	27 Jul 2027	264
Phase II Principal Mission			
Outbound	06 Dec 2028[1]	14 Aug 2029	251
Homebound	08 Nov 2030	17 Aug 2031	282

1. 2028 was chosen due to the lower risk of Solar flares.

Vehicle stages use liquid oxygen (LOX) and liquid hydrogen (LH$_2$) propellants, a mission design factor requiring long term storage facilities in LEO. The Earth departure stage (EDS) is reusable and is employed to boost phase I and phase II vehicles.

Mission parameters

Each vehicle comprises three stages. An EDS is used to achieve escape velocity, a Mars transfer stage (MTS) performs the braking maneuver into Mars orbit, and an Earth return stage (ERS) is used to escape from Mars orbit and to place the ERS on the Earth return trajectory.

On approach to Earth, a minor propulsive maneuver is required to acquire a 400 km orbit, whereupon Skylon approaches and conducts a rendezvous with the ERS to retrieve the crew and payload. Crewmembers then return to Earth in the Skylon, which lands at either Cape Canaveral or Kourou in French Guyana.

The locations of either Cape Canaveral or Kourou enable 28°.5 orbits to be achieved every three days. Since Skylon is capable of deploying 10,580 kg from Kourou and 10,790 kg from Canaveral in SSTO mode, a total of 1300 tonnes can be placed into a parking orbit in just twelve months.

The payload of each Precursor mission (Table 3.6) includes all surface equipment and the transfer ferries which provide a means of transport from the Martian surface to Mars orbit. With the exception of the ferries, the entire Precursor payload is deployed to the Martian surface and remotely activated, prior to crews embarking on the Principal mission. The ferries, which carry sufficient propellant for an abort to orbit, remain in Mars orbit and wait for the crew to arrive.

Table 3.6 Precursor and principal mission payloads [5]

Precursor Mission		Principal Mission	
Item	Mass (t)	Item	Mass (t)
Surface Habitat	30	Space Habitat	20
Nuclear power supply	10	2 capsules	2×4
Propellant factory	15	6 astronauts	0.45
2 trucks	2×5	7 spacesuits	1.54
Ferry (fuelled)	50	Consumables & water	43.7
Stores & equipment	19.8	Surface equipment & supplies	37.4
Assembly robot	2		

The primary mass of the Principal mission (Table 3.6) is the Space Habitat and the two aerobraking capsules. Other components of the payload include surface equipment and supplies which remain in Mars orbit until the crew arrive. In the event a landing is impossible, due to a dust storm for example, the crew survives using the payload until an Earth transfer window opens.

Lift mass and launch requirements

If three vehicles are used, the total lift mass for the Precursor mission totals 2346 t (Table 3.7) and 2234 t for the Principal mission. To achieve this, 273 flights are required for the Precursor mission and an additional 249 for the Principal mission, giving a total of 522 flights.

Cost and timescale

The total launch, vehicle, and manned element costs of Project Troy amount to
approximately US$68 billion (Table 3.8). The timescale for Skylon development and
the Troy Project is shown diagrammatically in Table 3.9.

Table 3.7 Lift mass and launch requirements [5]

Precursor Mission		Principal Mission	
	Lift Mass		Lift Mass
Liquid oxygen	510 t	Liquid oxygen	548 t
Liquid hydrogen	86 t	Liquid hydrogen	91 t
Stage related hardware	49.3 t	Stage related hardware	51.5 t
Mars payload	136.8 t	Stores, equipment & crew	91.3 t
Total Lift Mass	**782.1**	**Total Lift Mass**	**781.8**
Launch Requirements		Launch Requirements	
Item Lifted	**# of flights**	**Item Lifted**	**# of flights**
8 × EDS & MTS hydrogen tanks	8	2 × MTS hydrogen tanks	2
8 × EDS & MTS oxygen tanks	4	2 × MTS oxygen tanks	1
4 × SSMEs	2	ERS tanks	1
Mars equipment	15	1 × SSME + 2 × RL10-B2	1
Liquid oxygen	53	Payload & Mars equipment	10
Liquid hydrogen	9	Liquid oxygen & liquid hydrogen	68
Total Flights	**91**	**Total flights**	**83**

Table 3.8 Launch, vehicle and manned element costs [5]

Launch Costs Item	Cost US$ million
Skylon Development	11,705
4 production vehicles (800 flights each)	1,796
522 program launches	4,447
10 development launches	85
30 Operations Base launches	255
Total launch-related cost	*18,288*

Total Troy Vehicle Costs (Phase I + Phase II) Item	Cost US$ million
EDS/MTS module development	6,084
EDS/MTS module production	1,233
ERS development	1,247
ERS production (7 items)	145
Total propulsion hardware cost	**8,709**

Table 3.8, *cont.*

Manned Element Costs	
Item	Cost US$ million
Operations Base Station 10,000	
Mars Bases and Equipment	12,000
Space Payload and Equipment	5,000
Project Administration 15,000	
Total Manned Element Cost	**42,000**

Table 3.9 Timescale of Skylon development and the Troy Project [5]

Task ID	2010 to 2017	2017 to 2024	25	26	27	28	29	30	31	32
1										
2										
3										
4										
5										
6										
7										
8										
9										
10										
11										
12										
13										
14										

Key to Task IDs

1	Skylon development	6	Phase I depart	11	Phase II arrival
2	Skylon in Service	7	Phase I arrival	12	Surface exploration
3	Troy Mission	8	Establish base	12	Phase II return
4	Hardware development	9	Assemble Phase II	14	Phase II complete
5	Assemble Phase I	10	Phase II depart		

Mission requirements

Living space
The Project Troy team assumes a minimum gross volume of 60m^3 per crewmember. Given the Skylon payload volume of 202m^3, this enables a space habitat with a volume of ~180m^3 to be placed in LEO.

Life support
Nutritional requirements are based on a 75-kilogram person requiring 3000 kcal/day in space and 3500 kcal/day on the Martian surface. Life support requirements (Table

Table 3.10 Life support requirements (75 kg crewmember) [5]

Input	In Space (kg)	On Mars (kg)	Output	In Space (kg)	On Mars (kg)
Dry Food	0.660	**0.770**	Urine	1.522	**1.556**
Oxygen	0.917	**1.070**	Faeces	0.155	**0.182**
Water	2.631	**2.631**	Respiration/ perspiration	1.522	**1.556**
			CO_2	1.009	**1.177**
			Heat	3000 kcal	**3500 kcal**

3.10) were based on a six-crewmember mission departing 2028, spending 434 days in space and 451 days on the surface, resulting in a requirement for 4200 kilograms of food, 583 kilograms of oxygen and 3000 kilograms of water, assuming 500 kilograms of water per person is recycled.

Project Troy's environmental life support system utilizes lithium hydroxide to remove carbon dioxide produced by the crew, and relies on respiration being converted to water as a means to reduce the amount of water being transported to Mars.

Medical issues

The Project Troy team suggests providing an artificial gravity system to help crewmembers simulate reduced gravity during the outbound and inbound phases of the mission. The problems of a rotating system, as envisaged by the team, include perturbations as the crew move and also the effect of the Coriolis accelerations on the vestibular systems of the crew. In reality, it is likely the crew would resort to conventional countermeasures such as a vigorous exercise regime and pharmaceutical intervention strategies, such as those outlined in Chapter 8.

The issue of radiation is also addressed by the Project Troy team by distributing the 3000 kilograms of water in the ERS and the 9926 kilograms of the fuel cell reactants around a flare shelter to provide crews with adequate shielding in the event of a solar flare event.

Surface architecture

A nuclear-generated power plant is proposed as a fixed surface installation to manufacture propellants such as carbon monoxide and carbon suboxide (tricarbon dioxide) from the Martian atmosphere. The power plant would be landed together with the habitat and would require a mobile robot able to deploy, bury and commission the plant. Surface mobility would be achieved by tracked trucks similar to those used by the military.

For return trips from the Martian surface to Mars orbit, each expedition has a single stage ferry capable of transporting six crewmembers. In addition to providing the crew with an abort capability, the ferry can also carry payloads to orbit and return crews to the ERS for the return home.

Although the implementation of Project Troy would yield a new means of attaining orbit and perhaps drive down the cost of accessing LEO, viewed through the eyes of a mission planner, the design appears too awkward due to the number of flights required. Another reason Project Troy is unlikely to be realized in the near term, is that it demands the development of a new launch vehicle, always a fiscally painful and often prolonged process. Realistically, the architectures that have the best chance of success are those proposed by the international space agencies such as ESA and NASA.

EUROPEAN SPACE AGENCY

The ultimate objective of ESA's Aurora program is to land humans on Mars. While several mission architectures have been proposed, ESA has not chosen a reference scenario, arguing the level of technology improvement between now and when a manned mission can be realized precludes making such a decision.

Mission architecture

The mission architecture presented here represents a suggested mission profile based on studies conducted by ESA as a part of its Aurora program [4]. The architecture is a split-mission scenario in which elements of the architecture are deployed automatically, prior to the arrival of the crew. The crew size is restricted to four and the architecture is comprised of the key mission elements described in Table 3.11.

ESA's plan commences with one Ares V flight delivering the surface elements and the crew Mars ascent/descent vehicle to LEO. A second Ares V flight delivers a nuclear transfer stage that performs the TMI maneuver, sending the mission elements to LMO. The first phase of the human mission comprises two additional Ares V flights, the first delivering the Mars nuclear transfer stage to LEO, and the second delivering the Mars transfer habitation module and an additional transfer stage, providing the supplementary fuel for the MOI and TEI maneuvers. The second phase of the human mission is the launch of the crew into LEO onboard the Ares I. Once in LEO, the crew transfers to the transfer habitation module and return capsule.

ESA's plan represents a solid and robust architecture, utilizing existing or soon-to-be realized technology although, since much of this technology will developed by the United States, the plan assumes a collaborative effort with NASA, a decision that has yet to be finalized. The collaboration with NASA notwithstanding, ESA's plan, while certainly achievable in the timeframe promoted by the European agency, is still resource intensive, since each mission must expend several launch vehicles. Since expending launch vehicles incurs significant costs over the timeframe of conducting sustainable exploration, a better solution might be to develop an architecture that enables such exploration but without the need to discard hardware. Such an architecture is represented by the *Cycler* concept, described below.

Table 3.11 ESA Mars mission elements [4]

Transit Habitation Module		
Item	Value	Role
Overall Mass (t)	**23.0**	Sustains crew of four for up to two and a half years in
Consumables Mass (t)	**16.0**	deep space environment. Provides a storm shelter for
Pressurized Volume (m^3)	**300**	Solar Particle Events. Provides exercise facility for
Length (m)	**8.5**	crew and provides storage for consumables.
Main cylinder diameter (m)	**7.2**	

Surface Habitation Module		
Item	Value	Role
Overall Mass (t)	**32.0**	Sustains crew of four for up to 400 days on surface of
Habitable Volume (m^3)	**20.0**	Mars. Provides crew with storm shelter in case of Solar
Pressurized Volume (m^3)	**44.0**	Particle Events. Provides crew with airlock and four
Length (m)	**6.2**	suitlocks. Houses four ECLSS/EVA and five science
Diameter (m)	**6.2**	racks.

Pressurized Rover		
Item	Value	Role
Total Mass (t)	**9.6**	Sustains crew of two during sortie missions of fifteen
Pressurized Volume (m^3)	**49.0**	to twenty days on the surface of Mars.
Range (km)	**800**	
Cruising Speed (km/h)	**10.0**	

Mars Descent Stage		
Item	Value	Role
Total Mass (t)	**24.0**	Performs Mars descent from orbit and soft precision
Heat Shield diameter (m)	**8.8**	landing on Mars surface. Sustains crew during descent
Thrusters	**8**	to surface.
Height (m)	**6.4**	

Mars Ascent Vehicle		
Item	Value	Role
Total Mass (t)	**30.0**	Performs ascent from Mars surface and places crew in
Height (m)	**6.4**	a circular five hundred kilometer Mars orbit. Sustains
Diameter (m)	**7.0**	crew during ascent from surface.
Stages	**2**	

Mars Electrical Transfer Vehicle		
Item	Value	Role
Overall Length (m)	**16.0**	Delivers habitation module on Mars surface and
Width (m)	**7.0**	exploration tools (power plant, rover). Evaluates
Specific Impulse (s)	**2929**	potential application of Earth-Moon libration for
Payload mass delivered	**100.0**	mission to Mars and application of an electrical
to LMO (t)		spaceship powered by xenon thrusters.

Table 3.11 *cont.*

Mars Nuclear Transfer Stage		
Item	Value	Role
Total Mass (t)	**120.0**	Assures the propulsion required to conduct human
Diameter (m)	**10.0**	mission to Mars. Reduces mass required for insertion
Length (m)	**29.0**	into LEO. Requires development of enabling technol-
Specific Impulse (s)	**1000**	ogies such as cryo-cooler systems and nuclear thermal
Nuclear Engines	5	propulsion.

Mars Aerostationary Communication Orbiter		
Item	Value	Role
Launch Mass (kg)	**2700**	Ensures high data-rate communication relay to sup-
Lifetime (years)	**10**	port human activities in orbit and during surface
Communication link (Mbps)	**10**	missions. Continues to support robotic missions once
Circular equatorial orbit altitude (km)	**17030**	manned sorties have commenced.

GLOBAL AEROSPACE CORPORATION

Global Aerospace Corporation (GAC) has developed a pioneering architecture utilizing autonomous, solar-powered, xenon ion-powered spacecraft designed to orbit the Sun in cyclic orbits between Earth and Mars [8]. The spacecraft, which serve as astronaut hotels, or *Astrotels*, would continue their orbits and up to ten passengers would taxi between the Astrotels and planetary spaceports and surfaces, using shuttles.

Mars transit base
The GAC architecture envisions a Mars base, serving as an interplanetary transit location, populated by a crew complement of twenty tasked with exploration, science, resource surveys, life-cycle maintenance, propellant production and materials processing. Thanks to staggered crew rotations dictated by interplanetary transport orbit options, the Mars Base would be continually inhabited, ensuring an overlap between experienced and fresh personnel.

By maximizing in-situ resource utilization (ISRU), the Mars base is almost self-sufficient, requiring minimal resources from Earth. Although the Environmental Control and Life Support System (ECLSS) is not entirely closed, it is regenerative and life support gases and water would be extracted from the soil and atmosphere. In common with Zubrin's Mars Direct plan, the GAC architecture envisions the production of propellant in situ but to achieve this, and the other goals of the mission plan, requires a significant amount of hardware (Table 3.12).

Table 3.12 Global Aerospace Corporation Mars base mass summary[1]

Mars Base Systems	No. of Units	Unit Mass (t)	Total Mass (t)
Life Critical Systems			
Habitat	4	38.5	154.0
Washdown facility	2	0.9	1.8
Subtotal			**155.8**
Mission Support Systems			
120 kW Solar Array (100W/kg)	2	1.2	2.4
Power Management & Distribution	2	0.3	0.6
Energy Storage (NFRC packages)	2	1.0	2.1
Suit-up/Maintenance Facility	2	1.8	3.6
Pressurized Transporter	3	9.1	27.3
Open Rovers	3	1.0	3.0
Inflatable Shelter with Airlock	10	0.5	5.0
Communication Satellites	3	0.8	2.4
Crane	2	5.0	10.0
Trailer	2	2.0	4.0
Subtotal			**60.4**
Science and Exploration Systems			
Base Laboratory	2	13.6	27.2
Mobile Laboratory	3	9.1	27.3
200m Drill	1	2.3	2.3
10m Drill	3	0.1	0.3
UAV	3	0.3	0.9
Robotic Rovers	10	0.2	2.0
Weather Station	5	0.2	1.0
Survey Orbiters	2	0.8	1.6
Subtotal			**62.6**
Total			**278.8**

1. Adapted from 'Interplanetary Rapid Transit to Mars', by Kerry Nock, Angus McRonald, Paul Penzo, and Chris Wyszkowski. Paper Number: 03 ICES-2392. 2003.

Cycling orbits

The GAC architecture is based on what mission designers refer as *Cycler Orbits*, designed to enable sustained interplanetary transportation via regular encounters with Earth and Mars. A number of Cycler Orbit ideas have been proposed with the intent of supporting sustained Mars operations but, perhaps the most widely known is the concept devised by no other than Buzz Aldrin himself (Figure 3.7), who developed the Mars Cycler System (MCS) using techniques of orbital mechanics he developed at the Massachusetts Institute of Technology (MIT) during his Ph.D. studies.

Aldrin's MCS starts with a crew launch into LEO in a vehicle similar to Orion. In

Figure 3.7 Former astronaut, Buzz Aldrin, has his own proposal for mankind's next giant leap. The Cycler Orbit is based on reusable spacecraft that continuously cycle between Earth and Mars in permanent orbits. (Rick Stiles/United States Government.)

LEO, Orion docks with a Mars lander and a propulsion module previously launched from Earth. Once docked, Aldrin's three-element spacecraft performs a burn, placing it in a highly elliptical six-day orbit around the Earth, taking it almost halfway to the Moon. During the orbit, the spacecraft docks with a cargo vehicle carrying liquid oxygen (LOX) and hydrogen manufactured on the Moon. Once propellant is transferred to the propulsion module, the vehicle performs a TMI burn, placing it on a trajectory to rendezvous with the Cycler vehicle comprising an AstroTel, ion engines and a solar array (Figure 3.8). After conducting a rendezvous with the Cycler, Orion and the Mars lander separate from each other and dock with the Cycler and the crew transfers from Orion to the Cycler's habitation module. The crew then kicks back, unstows their entertainment supplies and prepares for a five-month voyage to Mars. Thanks to the Cycler's spinning habitation module simulating one-third Earth's gravity, the crew doesn't even have to worry too much about exercise. As the Cycler approaches Mars, the crew transfers back into Orion with the lander still attached and conducts an aerobraking maneuver to reduce speed. Once the aerobraking has been performed, the Orion settles into a low Mars orbit (LMO), the crew transfers to the lander, undocks from Orion, fires the lander's retrorockets and descends to the surface expecting a welcome reception from the incumbent crew, who has been working on Mars for more than two years.

The return trip is achieved using a Semi-Cycler, a hybridized Cycler designed to shuttle between Earth and Mars in a gravity-assisted orbit. The reason for using the Semi-Cycler is the very high velocity of the Cycler on arrival at Mars, requiring

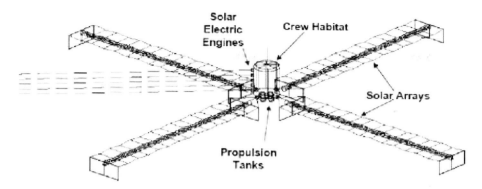

Figure 3.8 An AstroTel will perform cyclic orbits between Earth and Mars and will use solar electric propulsion for orbit corrections. (Global Aerospace Corporation/NASA Institute for Advanced Concepts.)

Orion to make a huge propulsive burn to catch up. Whereas the Cycler arrives at Mars traveling at 27,000 km/h, the Semi-Cycler orbits the Red Planet at only 8,000 km/h. For the return trip, the crew uses the lander of the incoming crew but, before they leave the surface of Mars, the crew needs to ensure the Cycler has enough fuel to get back to Earth. To do this the crew waits until the Cycler arrives in LMO and then launches an unmanned rocket filled with fuel to top up the Cycler's fuel tanks. Once this is accomplished, the crew boards the lander of the incoming crew and launches into LMO, where they dock with the Cycler which is joined with Orion. A TEI burn is then performed, sending the crew en-route for Earth. During this trip there is no artificial gravity, so the crew needs to spend a few hours a day exercising to maintain bone and muscle strength. As Orion approaches Earth, it undocks from the Semi-Cycler and conducts an aerobraking maneuver into the Earth's atmosphere, while the Semi-Cycler continues on its return trajectory. Landing is achieved by deploying recovery chutes and descending for an ocean splashdown.

The Cycler concept represents a flexible architecture, enabling options such as Stopover Cyclers which are direct transfers from Earth to Mars. Although the Stopover Cyclers entail a high-thrust propulsive maneuver at each end of the trajectory, the departure calls for a relatively low propulsive burn, since the velocity required is comparatively slow. The architecture's only real constraint is the requirement to launch on time with little margin for error, since the Cyclers are unable to make any changes in velocity. However, the GAC architecture resolves this problem with the use of transit stations and spaceports.

Transit stations
The problem of an Orion-Cycler rendezvous is the large amount of energy required to perform the burns necessary to effect what mission planners refer to as *plane changes*. Plane changes involve altering the inclination of an orbiting body's orbit, requiring significant fuel expenditure. Rather than waste propellant on effecting a plane change in this manner, the GAC architecture utilizes a spaceport located in a

high orbit where plane changes can be made with much less fuel because orbital velocities from high orbit locations are lower. In this mission plan, an Orion departs Earth, conducts a rendezvous and docking with a high-orbiting spaceport, whereupon the spaceport positions itself for the arrival of a Cycler. As the Cycler approaches, a shuttle vehicle transports crew and cargo from the spaceport to the Cycler/Astrotel. On arriving at Mars, the crew transfers to a Mars Spaceport, located near Phobos, and is ferried to the Mars Base onboard a Mars Shuttle.

Testing the plan
The GAC team has tested its architecture using computerized models to generate mission options, perform trade studies, analyze life support systems and calculate life cycle costs. Costs of the architecture were also calculated based on cost references of previous manned missions and parametric data developed by NASA. Also considered were more than one hundred sub-elements of the architecture and the various permutations of the plan. After much number-crunching, the GAC team calculated that, for a baseline Aldrin Cycler mission plan, the life-cycle cost would be about $117 billion, or about $13 billion more than the anticipated cost of returning to the Moon under the Constellation Program. Much of this cost ($69 billion) would be for flight development but, once the system was running, it would require an annual operating cost of just $2.8 billion, a figure that is twenty percent of the current NASA budget.

Cycler analyzed
Perhaps the greatest advantage of Cycler over many of the other architectures discussed in this chapter is its evolutionary approach to Mars exploration. Rather than embarking upon a 'flags and footprints' approach, the Cycler plan envisions concepts enabling a permanent inhabitation of Mars based on near-term development technology. Nevertheless, while the architecture may ultimately be implemented in the long-term, in the short-term the most viable architecture is the one designed under the auspices of NASA's new Vision for Space Exploration (VSE).

NASA DESIGN REFERENCE MISSION

In January 2004, President George W. Bush announced the VSE, directing NASA to return humans to the Moon by 2020 and to prepare for human exploration of Mars. As a part of the VSE, NASA will retire the Space Shuttle by 2010 and construct and fly the Crew Exploration Vehicle, now known as Orion, by 2014. To implement the VSE, NASA created the Exploration Systems Mission Directorate (ESMD) to lead the development of the required exploration systems. In 2005, the new NASA Administrator, Dr. Michael Griffin, restructured NASA's Exploration Program and created the Exploration Systems Architecture Study (ESAS) team, commissioned to provide definition of crew configurations and cargo launch systems to support the lunar and Mars exploration programs. The ESAS team examined hundreds of

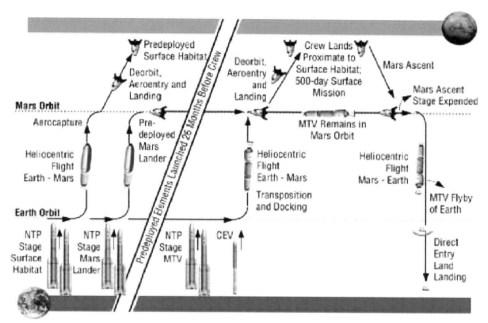

Figure 3.9 NASA's Mars Design Reference Mission (DRM) is a split mission concept. Cargo is transported to Mars in manageable units and checked out in advance of committing crews to their mission. Most of the cargo is sent on a minimum energy trajectory, while the manned mission tales advantage of a faster, more energetic trajectory. (NASA.)

different combinations of launch vehicles to perform the various Design Reference Missions (DRMs) to the Moon and Mars. The team also conducted trade studies examining different approaches to lunar and Martian sortie and outpost missions. Eventually, a series of DRMs were defined, including mission plans for transporting crews to and from the ISS, to the lunar surface and to Mars.

NASA's Mars Exploration DRM (Figure 3.9, Table 3.13) utilizes a conjunction class trajectory to minimize the exposure of crew to deep-space radiation and zero gravity, while also ensuring a long stay on the surface, thereby maximizing the science return of the mission [7].

For the most part, NASA's architecture utilizes current or soon-to-be realized launch vehicles and technologies and probably represents the most achievable robust mission plan described so far in this chapter. Some of the hardware required for the architecture is already being constructed, in preparation for a suborbital test launch of the Ares I in 2009, and in preparation for lunar missions sometime in the 2019 to 2020 timeframe. However, in common with ESA's architecture, NASA has yet to define in detail the various mission elements and surface exploration architecture, which is not surprising given the agency's current focus on transitioning from the Space Shuttle to Ares, and then returning humans to the Moon. However, a detailed architecture that follows the NASA mission plan has been proposed by

Table 3.13 NASA Mars Design Reference Mission stage by stage [7]

Stage	Timeframe	Description
1	December 2028	Cargo Lander delivered to LEO by Ares V. Multi-burn injection used at perigee to inject vehicles towards Mars.
2	January 2029	Surface Habitat delivered to LEO by Ares V. Multi-burn injection used at perigee to inject vehicles towards Mars.
3	October 2029	Cargo Lander aerocaptures into Mars orbit. Conducts de-orbit, aeroentry and landing.
4	November 2029	Surface Habitat aerocaptures into Mars orbit. Conducts de-orbit, aeroentry and landing. Surface Habitat deploys.
5	December 2030	Habitat and lander delivered to LEO by Ares V. Multi-burn injection used at perigee to inject vehicles towards Mars.
6	December 2030	Mars Transfer Vehicle (MTV) delivered to LEO by Ares I. Multi-burn injection used at perigee to inject vehicle towards Mars.
7	January 2033	Mars Ascent Stage delivers crew to low Mars orbit (LMO). Crew return to earth in MTV.
8	March 2033	Second crew departs Earth
9	July 2033	First crew arrives in LEO. Conduct direct entry land landing on Earth.
10	December 2033	Second crew arrives in LMO.

Total Mission Duration: 892-945 days

SpaceWorks Engineering Inc., (SEI). Not only does SEI's architecture utilize NASA's new family of launch vehicles, it also makes use of the TransHab technology, abandoned in 2000 and recently adopted by Bigelow Aerospace.

SPACEWORKS ENGINEERING INC. (SEI)

The architecture (Figure 3.10) proposed by SEI of Atlanta [10] probably represents the most innovative mission plan for reaching Mars, since it not only incorporates flight hardware currently being designed and developed by NASA but also utilizes the aforementioned TransHab technology. The lunar flight hardware, including a new family of launch vehicles and a new crew vehicle, is also being designed for the manned mission to Mars outlined in President George W. Bush's U.S. VSE. In conducting their study, SEI assumed the existence of the Orion crew exploration vehicle, the Ares I crew launch vehicle and the Ares V cargo launch vehicle. SEI also assumed environmental life support system technologies designed to support the Mars exploration campaign had been developed and made certain assumptions concerning the mission architecture (Table 3.14).

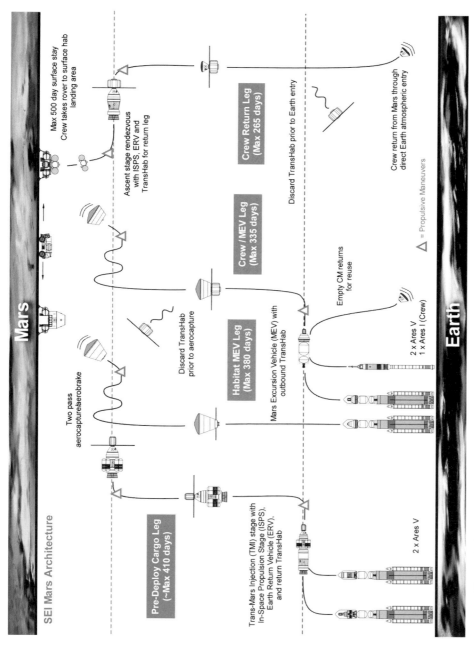

Figure 3.10 SpaceWorks Engineering Inc. Mars mission architecture. (SpaceWorks Engineering, Inc. (SEI).

Table 3.14 SpaceWorks Engineering Inc. architecture assumptions [10]

Assumption	Assumption Rationale
Conjunction Class Mission	Long-stay approach yields lower energy transfers from Earth to Mars. Penalties of conjunction class include large crew consumables and increased radiation exposure.
Crew Size	Three crewmembers considered adequate to accomplish flight, science and vehicle-maintenance duties during mission.
Sortie Missions.	No permanent base is established. Each mission visits a different site and is considered to be self-contained.
No NTR[1.] Engines for Transfer Vehicles.	Transfer, descent and ascent stages will use conventional chemical propellants and rocket engines. Despite advances in NTR technology it is assumed political forces will favor non-nuclear vehicles.
No In-Situ Propellant.	In-Situ Resource Utilization (ISRU) is eliminated, thereby removing development and operational risks associated with the technique.
Aerobraking	Aerocapture and aerobraking maneuvers are assumed to have been proven by 2030, using rigid aerobraking structures.
Zero-boiloff Cryogenic Storage.	A chemical propellant architecture will require capability to retain and use propellants transported from Earth. By 2030 it is anticipated boil-off of cryogenic hydrogen and oxygen is minimal through application of advanced cryocoolers.
No cryogenic propellant transfer	No orbiting refueling capability, permitting the transfer of propellants between stages once in orbit, is assumed.
Mars C&N[2.] Assets Exist	Spacecraft constellations have been placed in Mars orbit to serve as navigation aids and communication relays in support of human missions.

1. Nuclear Thermal Rocket.
2. Communications and Navigation.

Mission architecture

The architecture comprises a conjunction-class sortie mission requiring separate cargo and crew transfer phases, summarized in Table 3.15.

Table 3.15 SpaceWorks Engineering Inc. mission architecture [10]

Sequence	Vehicle/Module	Description
Cargo Transfer Phase		
1	Ares V Launch #1	Places Trans-Mars Injection (TMI) stage in LEO.
2	Ares V Launch #2	Places In-Space Propulsion Stage (ISPS), Mars-Earth leg return TransHab and Earth-Return Vehicle in LEO.

Table 3.15 *cont.*

Rendezvous		
3	TMI/ISPS/TransHab/ERV	TMI stage and ISPS/TransHab/ERV stack rendezvous in LEO and wait for Earth-Mars departure window.
Trans-Mars Injection		
4	ISPS/TransHab/ERV	Following TMI, the ISPS/TransHab/ERV stack undocks from spent TMI stage. Solar arrays on TransHab provide power to itself, ISPS and ERV.
Mars Approach		
5	ISPS/TransHab/ERV	Mars Orbit Insertion burn performed by ISPS stage. ISPS/TransHab/ERV stack remains in Low Mars Orbit (LMO) until required for crew's return trip.
Crew Transfer Phase		
6	Ares V Launch #1	Places Mars Excursion Vehicle (MEV), containing Martian surface habitat and descent stage, directly on a TMI trajectory.
7	Ares V Launch #2	Places another MEV (MEV2) and crew's Earth-Mars outbound TransHab into LEO. MEV contains ascent stage with minimal crew habitat, descent stage and pressurized rover.
8	Ares I Launch	Crew Exploration Vehicle (CEV) and crew delivered to LEO. During transit to Mars, crew lives in inflated TransHab.
Mars Arrival Phase		
9	MEV	Aerobraking/aerocapture maneuver conducted to place MEV in circular LMO.
10	Descent stage/surface habitat	Terminal descent maneuver performed, resulting in descent stage and habitat landing on Mars.
11	MEV2	Crew transfers to MEV2 ascent stage habitat. Transhab jettisoned. Crew land on Mars and drive to pre-deployed MEV.
Mars Departure Phase		
12	Ascent stage	Ascent stage of MEV lifts off from Mars and performs a Mars Orbit Rendezvous (MOR) with TransHab, ISPS and ERV.
13	TransHab/ISP/ERV	Trans-Earth Injection maneuver performed.
Earth Arrival Phase		
14	ERV/TransHab/ISPS	Crew transfers to ERV. TransHab and ISPS jettisoned.
15	ERV	Performs direct Earth entry.

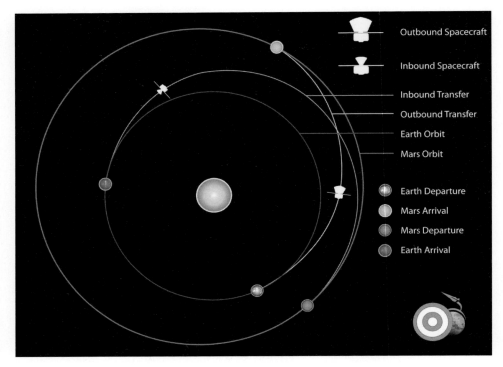

Outbound Spacecraft

Inbound Spacecraft

Inbound Transfer

Outbound Transfer

Earth Orbit

Mars Orbit

Earth Departure

Mars Arrival

Mars Departure

Earth Arrival

Figure 3.11 SpaceWorks Engineering Inc. Bullseye trajectory planning software. (SpaceWorks Engineering, Inc. (SEI).

Mission parameters

To calculate Earth-Mars and Mars-Earth transfer trajectories, the SEI team used its own interplanetary computer program known as Bullseye, an engineering analysis program that calculates interplanetary trajectories between two planets for a given time of flight (Figure 3.11).

Architecture flight hardware

Crew launch vehicle

The Ares I crew launch vehicle (Figure 3.12) consists of a five-segment reusable solid rocket booster (RSRB) first stage, a liquid oxygen/liquid hydrogen upper stage powered by a J-2X engine, and the Orion spacecraft (Figure 3.13), capable of carrying six astronauts.

Cargo launch vehicle

Capable of delivering as much as one hundred and thirty tonnes to LEO, the Ares V will be the largest and most powerful heavy lift vehicle ever built. Currently undergoing design and development, much of the Ares V hardware is Shuttle and Saturn-derived. Comprising a central liquid oxygen/liquid hydrogen core stage that

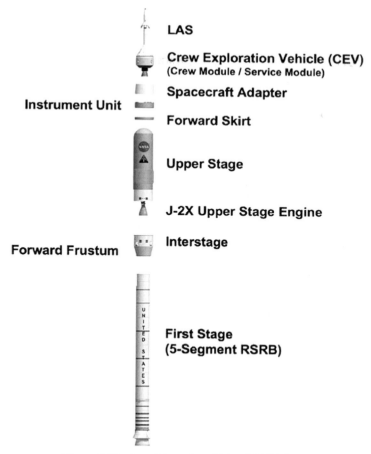

LAS

Crew Exploration Vehicle (CEV)
(Crew Module / Service Module)

Spacecraft Adapter

Instrument Unit

Forward Skirt

Upper Stage

J-2X Upper Stage Engine

Forward Frustum

Interstage

First Stage
(5-Segment RSRB)

Figure 3.12 Ares I launch vehicle. (NASA.)

Figure 3.13 SpaceWorks Engineering Inc. Earth Return Vehicle. (SpaceWorks Engineering, Inc. (SEI).)

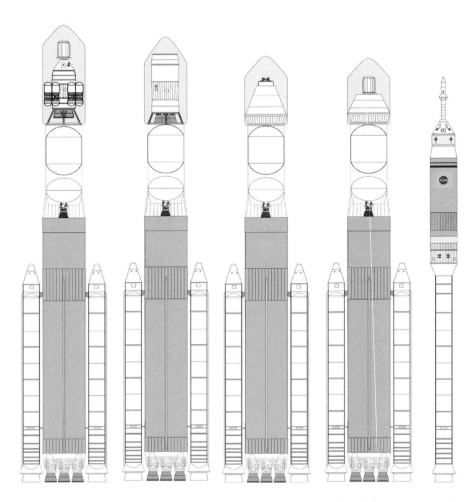

Figure 3.14 SpaceWorks Engineering Inc. launch vehicles showing flight elements under payload shroud. (SpaceWorks Engineering, Inc. (SEI).)

is basically a modified Shuttle external tank (ET) powered by five RS-68 engines, the Ares V design also incorporates two five-stage RSRBs that flank the core stage, and an Earth Departure Stage (EDS) powered by a J-2X engine (Figure 3.14).

Trans-Mars injection stage

The trans-Mars injection (TMI) stage (Figure 3.15) comprises two propellant tanks, an intertank adaptor, a payload adaptor, a thrust structure, four RL10B-2 engines and power subsystems. Designed to place the ISPS, return TransHab and ERV on a direct Mars transfer trajectory, the TMI is placed into LEO by a single Ares V launch.

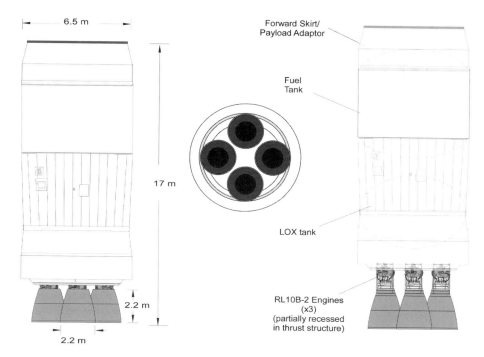

Figure 3.15 SpaceWorks Engineering Inc. trans-Mars injection stage. (SpaceWorks Engineering, Inc. (SEI).)

In-space propulsion stage
The in-space propulsion stage (ISPS) core (Figure 3.16) comprises two propellant tanks, an intertank adaptor, a payload adaptor, a thrust structure, three RL10B-2 engines and power subsystems. Attached circumferentially to the core are six sets of cylindrical drop-tanks, each consisting of a fuel and oxidizer tank and an intertank adaptor. The purpose of the ISPS is to provide the MOI burn for the cargo phase of the mission and the TEI for the crewed return phase. MOI is achieved by using the drop tanks, which are jettisoned after completion of the maneuver.

In-space transfer habitats
With the exception of the amount of stored crew consumables the two TransHabs used to support the crew are identical. Used to support the crew during the outbound and inbound Earth-Mars/Mars-Earth transits, the TransHabs (Figure 3.17) are rugged inflatable habitats consisting of Multi-Layer Insulation (MLI) micrometeorite protection and redundant inflatable bladders, which, when inflated, provide sixty cubic meters of habitable volume.

Power to the TransHabs is provided by four solar arrays, while thermal control inside the habitats is maintained by a two-fluid/Freon active system in which water is pumped throughout the central core and via coldplates attached to all major

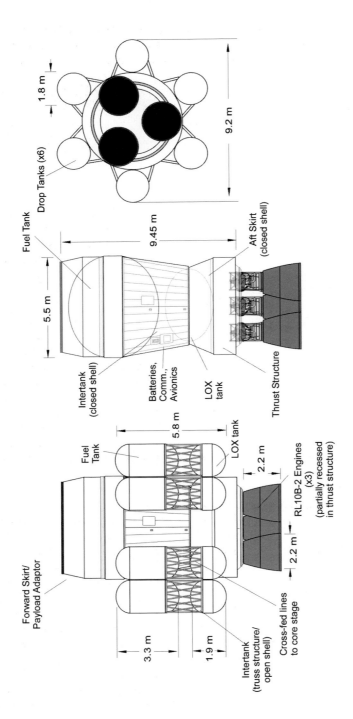

Figure 3.16 SpaceWorks Engineering Inc. in-space propulsion stage. (SpaceWorks Engineering, Inc. (SEI).)

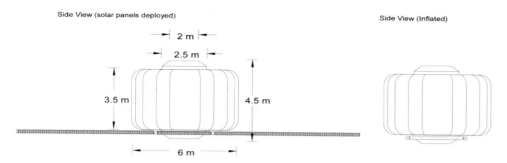

Figure 3.17 SpaceWorks Engineering Inc. TransHab. (SpaceWorks Engineering, Inc. (SEI).)

equipment. Using this system, water carries the heat to the top and bottom edges of the habitat, where it is transferred to the Freon loop via heat exchangers.

The TransHab's life support system is characterized by an arrangement in which water is filtered through multifiltration, waste water is reclaimed through vapor compression distillation and solid waste is vented to space. To remove carbon dioxide generated by the crew, the habitat is fitted with a molecular sieve. Once the carbon dioxide is combined with tanked hydrogen to produce CH_4 and water in a Sabatier reactor, the CH_4 is vented to space, the water is converted to oxygen and hydrogen via electrolysis using an oxygen generator, the hydrogen is routed back into the Sabatier reactor, and the oxygen is fed back into the cabin atmosphere.

The design of the TransHab was accomplished using *HabSizer* [10], a parametric software tool developed by SEI. HabSizer enabled the SEI team to rapidly size the habitation elements and perform trade studies on the elements.

Mars Excursion Vehicle elements
SEI's architecture requires two Mars Excursion Vehicles (MEVs) to enter the Martian atmosphere and land on the Red Planet. Each MEV comprises a propulsive descent stage, a payload housed within an outer heat shield, an ascent stage and a pressurized rover (Figure 3.18).

Design of the heat shield was accomplished using *Sentry* [2], a thermal protection system (TPS) sizing program developed by SEI. The program also enabled SEI to perform trade studies of candidate ablative TPS materials by analyzing factors such as heat transfer and radiative effects. Accordingly, SEI determined the best material option, based on mass efficiency and low technology risk, was Phenolic Impregnated Ceramic Ablator (PICA).

The design of the descent stage is based on an open struss structure, housing four rocket engines, propellant tanks and subsystems, an arrangement providing clearance for docking and transport of the pressurized rover. The LOX/LH2, used to fuel the engines as the MEV descends to the surface, is stored in spherical tanks arranged circumferentially of a circular platform. While the power for the (uncrewed) surface habitat MEV during its entry, descent and landing (EDL) is

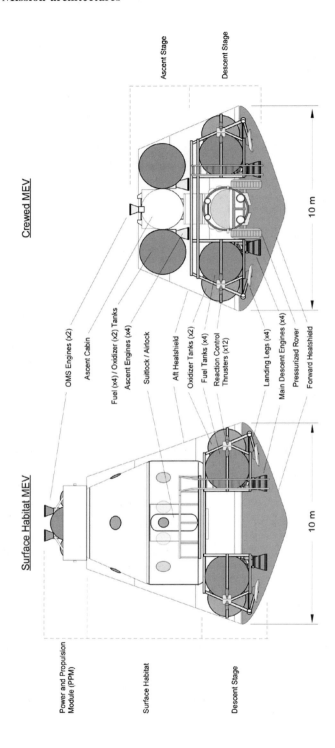

Figure 3.18 SpaceWorks Engineering Inc. Mars Excursion Vehicle. (SpaceWorks Engineering, Inc. (SEI).)

supplied by batteries, the electrical power for the crewed MEV during EDL is provided by an Advanced Stirling Radioisotope Generator (ASRG) [3].

The ascent stage transports the crew from the Martian surface to LMO. During the short ride to orbit, the crew is accommodated inside a pressurized cylinder similar to an aircraft cockpit. Fitted with four rocket engines, a cluster of reaction control thrusters and an orbital maneuvering system (OMS), the ascent stage must not only provide the propulsion to place the crew in LMO but also perform post-aerocapture orbit circularization and entry initiation burns. In the event of an incident delaying ascent and rendezvous, the stage is equipped with contingency supplies and consumables.

Entry, descent, landing and Mars ascent
The EDL and Mars ascent concept of operations (Table 3.16) was defined using a Program to Optimize Simulated Trajectories (POST)-II.

Architecture surface hardware

Mars Surface Habitat
The Mars Surface Habitat (MSH), which will support the crew during their surface stay, is a two-level rigid pressure vessel providing eighty-four cubic meters of habitable volume. The upper level is divided into three separate areas for personal crew quarters, while the lower level provides the crew with a science laboratory, a medical station and hygiene facilities. Access to the MSH is via a hatch in the lower level floor connecting to the pressurized rover, while access to the surface is via two airlocks, also on the lower level.

Table 3.16 Entry, descent, landing and Mars ascent [10]

Sequence	Maneuver	Description
1	Aerocapture	Inbound MEV is decelerated from arrival velocity of ~ 7 kms to Mars orbital velocity using aerodynamic forces.
2	Aerobraking	Performed to reduce energy of MEV's orbit.
3	Mars entry	(i) MEV decelerates aerodynamically to 10km altitude. (ii) Fore and aft heat shields jettisoned. (iii) Liquid rocket engines on descent stage ignited.
4	Terminal descent Phase	Propulsive deceleration and aerodynamic drag results in touchdown at vertical velocity of 2 m/s.
5	Ascent to orbit	Ascent stage engines ignited lifting stage to 100 by 400 km orbit.
6	Circularization burn	Performed at apoapsis in preparation for rendezvous with orbiting ISPS, return TransHab and ERV.

Power is supplied by one of two ASRGs, one of which is kept in reserve in the event of a contingency, while the MSH's thermal control and ECLSS systems are almost identical to those fitted in the TransHabs. Crew accommodations are also similar to those provided by the TransHabs but are modified to take into account the one third Martian gravity. Of interest is the absence of resistive exercise equipment, since the SEI team rationalized sufficient exercise would be undertaken by crews in the performance of extensive EVA operations. Also absent is a radiation shelter, a feature the SEI team considered unnecessary, arguing the Martian atmosphere provides sufficient protection from SPEs [10].

Habitat egress is via an airlock design featuring rear entry EVA suits docked to the sides of the interior of the airlock, an arrangement that not only permits crews regular access to the Martian surface but also prevents contamination of the internal habitat environment.

Pressurized rover

When the crew arrives on the Martian surface, it brings with it a four-wheeled pressurized ASRG-powered rover (Figure 3.19) capable of transporting all three crewmembers. Shortly after landing, the astronauts board the rover and drive to the previously deployed surface habitat. Upon completion of surface operations, the crew drives back to the crewed MEV.

Figure 3.19 SpaceWorks Engineering Inc. rover. (SpaceWorks Engineering, Inc. (SEI).)

Architecture masses

To define the masses of each architecture element, the SEI team utilized industry standard mass estimating relationships for each subsystem of each component, a process that generated sizing models, enabling the team to accurately estimate the payload for each launch.

While the first Ares V cargo ETO launch approaches the maximum LEO payload capability of the Ares launch vehicle, the payloads delivered to LEO for the crew launches is much lower than the maximum Ares V LEO payload. This is because the crew delivery launches must place payloads on a Mars transfer trajectory, which requires greater energy.

Mars exploration campaign

Predicting when demonstration flights supporting a manned mission to Mars might occur can only be based on the existing VSE Mars Exploration Architecture, which predicts such a mission taking place in 2031.

Similarly, predicting the cost of such a mission must be based on the assumption that NASA will return to the Moon and that the launch vehicles supporting the lunar return are ready to be incorporated into the human Mars exploration architecture (Table 3.17).

Table 3.17 Life cycle cost estimate for human Mars exploration architecture[1] [10]

Cost Item	LCC (M$) (FY2007)	Cost Item	LCC (M$) (FY2007)
Trans-Mars Injection Stage	4,630	Operations (Mission, EVA)	9,430
In-Space Propulsion Stage	3,930	Facilities: Launch	450
Crewed MEV Ascent Stage	7,580	Facilities: Mission Operations	2,450
Crewed MEV Descent Stage	4,450	Surface Systems	1,130
Pressurized Rover	2,970	Launch Vehicles	14,130
Surface Habitat MEV Propulsion Module	510	Technology Development	1,000
Surface Habitat MEV Descent Stage	4,220	Robotic Missions	1,000
TransHabs	6,930	Program Integration	9,020
Earth Return Vehicle	510	Reserves	12,630

Total $ 96.810 billion

1. Estimate accounts for all major components required in realizing lunar VSE objectives.

Mission risk

Loss of mission (LOM) and loss of crew (LOC) risks for the SEI mission architecture were assessed using a fault tree and event sequence diagram approach described in NASA's Probabilistic Risk Assessment Procedures Guide [11]. An analysis of risks for the architecture revealed a LOM would occur every 2.6 missions and LOC to occur once every 8.5 missions [11]. The primary contributors to such high LOM and

LOC values included the high risk associated with catastrophic failure during the Mars EDL phase and the long periods in deep space. Other, secondary contributors included the risk associated with an Ares V launch, each one calculated to have a three percent probability of failure. Since the architecture requires four such launches, the Ares V obviously represents a significant failure probability. While these numbers may seem high, as systems become more refined and developed between now and a potential launch date in 2031, the chances of LOM and LOC will decrease to five percent or less.

Give the innovative features of the SEI architecture and its use of hardware NASA intends to use on its own version of a manned mission to Mars, it is the SEI plan that will be referred to as the baseline mission throughout the remaining chapters. However, before moving on to examine the hardware needed to realize the SEI/NASA mission, it would be remiss to avoid discussing an architecture that has spawned hundreds of pages of discussion in the space blogosphere.

DIRECT 2.0

If you meander through the recesses of the space blogosphere these days, it is likely you will quickly find yourself immersed in a community in emotional upheaval. The source of much of the turmoil is the space community's dissatisfaction with NASA's current architecture for implementing the VSE, a ruckus that has resulted in some space supporters offering their own visions of "how to do it right". Some of the plans propounded in the blogosphere may be considered elaborate, others are more science fiction than science, while some, like DIRECT, have generated heated discussion and even some initial embarrassment within NASA, since the plan was conceived mostly by those still working for the agency! In this section we examine the architecture that has resulted in controversy and some consternation.

DIRECT [6] was published in October 2006, by a team led by Ross Tierney. The study was designed to persuade NASA to consider replacing the planned Ares I and Ares V with a single launch vehicle developed directly from existing Space Shuttle components [6]. The DIRECT launch vehicle would be based entirely on existing Space Shuttle hardware and infrastructure and designed to be configured in several different ways to match specific mission profiles.

On October 25, 2006, the study was submitted to NASA Administrator, Michael Griffin and shortly thereafter, NASA provided a critique, resulting in the re-evaluation of the entire proposal in a refinement study. After several months of revised calculations, feedback and critical analysis, Tierney and his team, which included many NASA engineers and mid-level managers, published DIRECT v2.0 on May 10th [6]. If adopted, DIRECT v2.0 promises to save NASA $35 billion over the next twenty years.

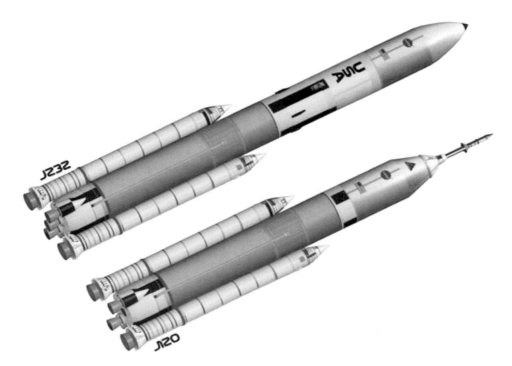

Figure 3.20 DIRECT 2.0 Jupiter 120 and 232 launch vehicles. (António H.F. Maia/ DIRECT.)

Jupiter launch system

Jupiter-120 and Jupiter-232 overview
The hardware (Figure 3.20) for DIRECT's Jupiter-120 and Jupiter-232 launch vehicles consists of many elements familiar to those who have followed the Space Shuttle missions over the last three decades (Table 3.18).

The Jupiter launch vehicles
The elements depicted in Figure 3.20 and described in Table 3.18 constitute the Jupiter Launch System (JLS). The crew variant of the Jupiter-120 comprises one cryogenic stage, two main engines, and no upper stage and will be capable of delivering 45 tonnes and Orion into LEO, whereas the cargo variant will be capable of launching 48 tonnes into LEO. This launch capability compares favorably with the Ares I lift-capacity of 22 tonnes into LEO, effectively duplicating the performance of existing launch vehicles such as the Delta IV Heavy and Atlas V.

The second launch vehicle, designated Jupiter-232, comprises two cryogenic stages, the first fitted with three engines, and the upper with two engines. The Jupiter-232 is capable of delivering 108 tonnes to LEO.

Table 3.18 DIRECT Jupiter120 and 232 launch vehicle concept specifications [6]

	Jupiter-120		Jupiter-232
GLOW	2,033,940 kg	**GLOW**	2,339,490 kg
CLV Launch Abort System Mass	6,565 kg	CaLV Aero Fairing Mass	2,279 kg
Booster		**Booster**	
Propellants	PBAN	Propellants	PBAN
Usable Propellant	501,467 kg	Usable Propellant	501,567 kg
# Boosters/Type	2/4-segment Shuttle RSRM	#Boosters/Type	2/4-segment Shuttle RSRM
Core Stage		**Core Stage**	
Propellants	LOX/LH2	Propellants	LOX/LH2
Usable Propellant	728,002 kg	Usable Propellant	729,002 kg
# Engines/Type	2/RS-68	# Engines/Type	3/RS-68
Engine Thrust @100%	SL[1] 297,557 kgf Vac[2] 340,648 kgf	Engine Thrust @ 100%	SL 297,557 kgf Vac 340,648 kgf
Core Burn time	446.0 sec	Core Burn time	292.0 sec
LEO Delivery Orbit	77.5 × 222.2 km @ 28.5°	**Second Stage EDS**	
Maximum Payload (Gross)	46,635 kg	Propellants	LOX/LH2
Maximum Payload (NET)	41,971 kg	Usable Ascent Propellant	216,012 kg
[1.] Sea level.			
[2.] Vacuum.		#Engines/Type	2/J-2XD
		Engine Thrust @ 100%	124,057 kgf
		Ascent Burn Time	392.0 sec
		LEO delivery Orbit	55.6 × 222.2 km @ 28.5°
		Maximum Payload (Gross)	105,895 kg
		Maximum Payload (NET)	95,305 kg

Payload

NASA's plan calls for a capability to deliver 150 tonnes to LEO in two launches, a payload many space observers consider marginal to ensure the safety of six astronauts during a Mars mission. The 150 tonnes of Initial Mass in Low Earth Orbit (IMLEO) requirement is not a concern for DIRECT planners since the two-launch Jupiter-232 is capable of delivering 95 tonnes to LEO *per flight*, thereby exceeding NASA's IMLEO requirement by 40 tonnes!

Integration and utilization of Shuttle-derived technology

Solid rocket boosters
DIRECT intends to use the Space Shuttle 4-segment SRBs which, following their redesign in the wake of the Challenger disaster in January 1986, have performed flawlessly. Since the SRBs are existing, man-rated elements there is no development cost or schedule impact and, because there are sufficient SRB components to support more than a hundred Jupiter missions, there are also no costs involved. NASA's architecture also intends to use the SRBs in their new family of launch vehicles but, instead of a four-segment SRB, NASA's plan requires the development of a five-segment version using a different fuel formulation.

External tank
For the Jupiter Common Core Booster (CCB), DIRECT decided to use the Space Shuttle's ET. The CCB represents the principal element with which the other JLS components will be integrated and will serve as the building block for subsequent derivatives of the JLS. Once again, DIRECT avoids the disruption to production, transportation, and workforce infrastructure by deciding to use the ET in its current configuration rather than build a different one, as mandated by NASA. To become the Jupiter CCB, the ET will require modifications, such as strengthening the sidewalls, designing a new avionics system, and developing some new plumbing.

Integration and utilization of existing technology
DIRECT does not use the SSMEs, arguing that, since the engines must be replaced after twenty uses, it doesn't make any sense to use them on a disposable Jupiter Common Core (JCC). Instead, DIRECT uses the Pratt & Whitney Rocketdyne RS-68 engine, originally designed for the US Air Force's Delta IV program. Inexpensive enough to be disposable and more than one and a half times as powerful as the SSME, the RS-68 requires only man-rating before being qualified for flight, thereby removing any significant development costs.

Mission architecture
Unlike all the other architectures described in this chapter, DIRECT proposes a series of missions to Near Earth Objects (NEOs), the moons of Mars and manned Mars orbital missions, before embarking upon the manned mission to the surface of Mars. DIRECT's argument is that pursuing such an incremental approach is necessary to demonstrate all the technologies required to land humans on Mars. Once the NEO and Mars orbital missions have been achieved, the elements of these precursor missions are combined into one coordinated mission to perform the first manned surface exploration of Mars.

DIRECT's manned mission architecture comprises six flight elements (Table 3.19) and just three launches. The first launch of the manned mission (Figure 3.21) is of a Jupiter-232 Cargo launch vehicle, which places the Mars in-situ resource utilization (ISRU) Lander and Crew Surface Ascent Vehicle (CSAV) in LEO. In LEO, these elements are refueled at a propellant depot deployed earlier, and then depart for

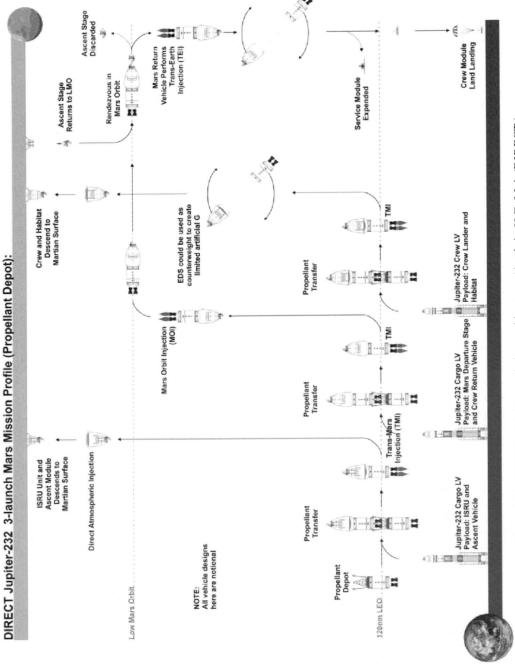

Figure 3.21 DIRECT 2.0 Mars mission architecture. (António H.F. Maia/DIRECT.)

Table 3.19 DIRECT flight elements [6]

LAUNCH #1. ISRU Lander and Crew Ascent Vehicle

Element	Mass (kg)	Element	Mass (kg)
Earth Entry/Mars Ascent Capsule	4,829	Inflatable Laboratory Module	3,100
Ascent Stage Dry Mass	4,069	15 kWe DIPS cart	1,500
ISRU Plant	3,941	Unpressurized Rover	550
Hydrogen Feedstock	5,420	3 Teleoperable Science Rovers	1,500
Keep-alive Power System	825	Water Storage Tank	150
150 kW Nuclear Power Plant	11,425	Science Equipment	1,770
Power Cables	837	Communication System	320
Total Cargo Mass 40,236 kg			
Vehicle Structure	3,186	Terminal Propulsion System	1,018
Total Landed Mass 44,440 kg			
Propellant	10,985	Forward Aeroshell	9,918
Parachutes & Mechanisms	700		
Total Payload Mass 66,043 kg			

LAUNCH #2. Crew In-Space Earth Return Vehicle

Element	Mass (kg)	Element	Mass (kg)
Habitat Element	26,581	30 kW Power System	3,249
Life Support System	4,661	Thermal Control System	560
Crew Accommodation & Consumables	12,068	Structure	5,500
EVA Equipment	243	Science Equipment	600
Information Management	320	Spares	1,924
Total Cargo Mass 29,105			
TEI Stage Dry Mass	4,806	Earth Return RCS Propellant	1,115
Propellant Mass	28,866	Aerobrake	10,180
Total Payload Mass 74,072			

LAUNCH #3. Crew Outbound and Mars Surface Habitat

Element	Mass (kg)	Element	Mass (kg)
Mars Surface Habitat	28,505	Spares	1,924
Life Support System	4,661	Thermal	550
Crew Accommodations	12,058	Structure	5,500
EVA Equipment	243	Crew	500
Information Management	320	Unpressurized Rover	550
Power	3,249	EVA Consumables	416
		EVA Suits	940
Total Cargo Mass 30,941			
Vehicle Structure	3,186	Terminal Propulsion System	1,018
Total Landed Mass 35,145 kg			
Propellant	11,381	Forward Aeroshell	13,580
Parachutes & Mechanisms	700		
Total Payload Mass 60,806 kg			

Mars. Once the Mars ISRU Lander and CSAV are in place on the Mars surface, the ISRU system begins to generate propellant required by the crew to depart the surface and rendezvous with the Earth return habitat in Mars orbit.

Once the propellant is generated, a second launch of a Jupiter-232 Cargo launch vehicle places the Mars Departure Stage (MDS) and Crew Return Vehicle (CRV) in LEO, where, following refueling, it departs for Mars. On arrival at Mars, the flight elements are inserted into Mars orbit and await the crew. A third launch of a Jupiter-232 crew launch vehicle places the Crew Lander and Crew Habitat in LEO. Once again, following insertion into LEO, propellant transfer is performed and the flight elements perform the TMI and depart for Mars. On arrival in Mars orbit, the crew and Habitat descend to the Martian surface to commence six hundred days of exploration and science activities. The return commences with the Ascent Stage returning to LMO where it performs a rendezvous and docking with the CRV. The Ascent Stage is then discarded and the CRV performs the TEI, placing the vehicle en-route to Earth.

From a manned Mars mission perspective, DIRECT scores points in its ability to deliver large payloads to LEO. The hardware suggested also utilizes more of the existing Space Shuttle infrastructure than NASA's current VSE plan. This means, if DIRECT were adopted, less time would be required to develop new systems, thereby closing the gap between lunar flights and Mars missions. However, despite the benefits of DIRECT, it is likely VSE will continue to move ahead, although perhaps not in its original configuration. Under the existing schedule, despite a new presidency in 2009, most of the current members of Congress who accepted NASA's Ares I/V concept will still be there when Ares I becomes operational, and when Ares V development begins. Since Congress will probably feel a sense of obligation to finish what it has started, it is probable Ares V will be built and it will be these launch vehicles that ultimately support a manned Mars mission.

IN SUMMARY

This chapter has provided a general overview of contrasting manned Mars mission architectures. Since the first launch of any architecture will probably not depart for twenty years, some elements described here will evolve while others will be discarded. Another possibility is the discovery of a ground-breaking propulsion technology that may render many of the architectures described here obsolete, a topic discussed in Chapter 5. However, regardless of which mission plan is eventually selected, the common architecture elements such as abort options and the challenges of entry, descent and landing (EDL) will still apply and these are the features described in the following chapter.

REFERENCES

1. Bonin, G. Reaching Mars for Less: The Reference Mission Design of the MarsDrive Consortium. 25th International Space Development Conference (ISDC), Los Angeles, California, 4–7 May, 2006.
2. Bradford, J.E.; Olds, J.R. Thermal Protection System Sizing and Selection for RLVs Using the Sentry Code. AIAA-2006-4605, 42nd AIAA/ASME/SAE/ASEE Joint Propulsion Conference and Exhibit, Sacramento, California, July 9–12, 2006.
3. Chan, J.; Wood, J.G.; Schreiber, J.G. Development of Advanced Stirling Radioisotope Generator for Space Exploration. NASA TM-2007-214805. 2007.
4. Integrated Exploration Architecture. Strategy and Architecture Office. European Space Agency. HM-HS/STU/TN/OM/2008-04002. Issue 1. 2008.
5. Martin, T (Editor). Project Troy. A Strategy for a Mission to Mars. Reaction Engines Limited, Abingdon, Oxfordshire, U.K., 25th January, 2007.
6. Metschan, S.L.; Longton, C.A.; Tierney, R.B.; Maia, A.H.F.; Metschan, P.J. Achieving the Vision for Space Exploration on Time and Within Budget. AIAA 2007-6231. AIAA SPACE 2007 Conference and Exposition, 18–20 September 2007, Long Beach, California.
7. NASA Exploration Systems Architecture Study. Final Report. NASA-TM-2005-214062, November 2005.
8. Nock, K.T. Cyclical Visits to Mars via Astronaut Hotels. Global Aerospace Corporation. Presentation to the NASA Institute for Advanced Concepts, 3rd Annual Meeting. June 5–6, 2001.
9. Simonsen, L.C., and Nealy, J.E. Mars Surface Radiation Exposure for Solar Maximum Conditions and 1989 Solar Proton Events. NASA TP-3300. 1993.
10. St. Germain, B.; Olds, J.R.; Bradford, J.; Charania, A.C.; DePasquale, D.; Schaffer, M., and Wallace, J. Utilizing Lunar Architecture Transportation Elements for Mars Exploration. SpaceWorks Engineering, Inc., Atlanta. AIAA SPACE 2007 Conference and Exposition, 18-20 September 2007, Long Beach, California.
11. Stamatalos, M. Probabalistic Risk Assessment Procedures Guide for NASA Managers and Practitioners. NASA Office of Safety and Mission Assurance, NASA Headquarters, Washington D.C., Version 1.1., Aug. 2002.
12. Von Braun, W. Project Mars: A Technical Tale. Collector's Guide Publishing Inc. December, 2006.
13. Wooster, P.D.; Braun, R.D.; Ahn, J.; Putnam, Z.R. Mission Design Options for Human Mars Missions. The International Journal of Mars Science and Exploration. Mars 3, 12–28, 2007.
14. Zubrin, R.M.; Baker, D.A.; Gwynne, O. Mars Direct: A Simple, Robust, and Cost Effective Architecture for the Space Exploration Initiative, AIAA-91-0328. 1991.

4

Abort modes and the challenges of entry, descent and landing

For those tasked with designing robust and safe mission architectures, a manned mission to Mars presents problems far exceeding the challenges that have been faced before. For example, one of the overriding factors making a Mars mission so fundamentally different from lunar missions is the lack of opportunity to abort the mission, since, once the trans-Mars injection maneuver (TMI) has been performed, the crew is committed to a four to six month journey. Another equally formidable challenge is assuring the safety of the crew during the nerve-wracking entry, descent and landing (EDL), a phase most planners agree is the most dangerous part of the entire mission.

In this chapter, the abort modes available to a crew en-route to Mars are reviewed. Also discussed are the technology and systems advances required to safely land humans on the Red Planet and the approaches being investigated to improve EDL technology.

ABORT OPTIONS

When astronauts return to the Moon within the next fifteen years they may take some comfort in the knowledge an escape route is always available from lunar orbit or the lunar surface in the event of a contingency. For crewmembers en-route to Mars, it will be a very different story.

While abort modes during the launch and low Earth orbit (LEO) phases of a lunar mission will also be available to crews embarked upon a Mars missions, the abort options once the vehicle has departed LEO will be extremely limited. For crews departing LEO for Mars on a mission following the SpaceWorks Engineering Inc. (SEI) architecture (see Chapter 3), which this book is using as its design reference mission (DRM), any chance of returning to Earth expires shortly after performing the TMI maneuver. The next opportunity the crew has for an abort is as the spacecraft approaches Mars, more than four months later. If a contingency event occurs at this stage, the crew may be able to perform what mission planners refer to as a free return trajectory, the details of which are described in the following section. If a problem develops while the crew is orbiting Mars, the only options available are to either remain in orbit or abort to the surface. Once on the surface, the crew is then committed to a five-hundred day stay and if they encounter a problem requiring a return to orbit, the consequences for the crew are potentially life-threatening. Not only would the crew be exposed to zero gravity for up to five hundred days, but the excessive radiation outside of the Martian atmosphere would inflict severe physiological damage and dramatically increase the risk of cancers among the crew. Furthermore, the psychological effects of being confined in a habitat the size of a school bus for such a length of time would be challenging, to say the least!

Free return trajectory
Although abort opportunities are available to crewmembers en-route to Mars, none ensures a rapid return to Earth. Many of the abort options are based on what mission planners refer to as a *free return trajectory*, a trajectory in which a spacecraft traveling away from a primary body such as the Earth is modified by the presence of a secondary body such as Mars, causing the vehicle to return to the primary body. To utilize this abort option, the spacecraft must enter a planet's orbit in order to take advantage of the planet's gravity, which provides the energy for the *free return*. Assuming the trajectory is performed correctly, the vehicle can return to the primary body without any mid-course corrections or maneuvers, enabling the spacecraft to travel a great distance without the use of any additional propellant, hence the term *free*. However, once it has returned to the primary body, the spacecraft must still use propellant to slow the craft into orbit and to effect a de-orbit burn for the return to the surface.

In terms of the length of time spent in space, the two-year free return trajectory is perhaps the most attractive to astronauts. In this free return trajectory variation, assuming the arrival at Mars is aborted, the outbound spacecraft would be inserted into a two-year orbit period, resulting in the vehicle completing one orbit about the Sun while the Earth completes two orbits about the Sun. This type of free return represents the shortest Earth return time of any of the free return trajectory options. However, although the two-year return option might be desirable in terms of time spent in space, for mission planners the trajectory's drawback is its requirement for a high Mars entry velocity, calling for an advanced aerocapture system (remember, the spacecraft must enter Mars orbit in order to take advantage of the planet's gravity to ensure the free return). However, this problem may be solved by employing a propulsive abort maneuver at Mars, which could lower the Mars entry velocity.

Perhaps the greatest challenge for mission planners in the event of an abort is the requirement to support the crew during the duration of their transit to Earth. Two years is about the same duration as a complete round-trip Mars mission, so consumables become an issue since, in the event of an abort, the crew would not have access to the in-situ consumables on the Martian surface (the architecture proposed by SEI takes this into consideration). Instead, the crew might have to implement emergency rationing, reduce water consumption and begin the process of converting propellants to consumables, such as water and oxygen.

CHALLENGES OF ENTRY, DESCENT AND LANDING

Assuming the crew enjoys a problem-free deep space transit from Earth to Mars, the next challenge awaiting them as they approach the Red Planet is performing the nail-biting EDL sequence which begins half an hour prior to touchdown. The EDL sequence (Table 4.1) includes phases such as cruise stage separation, parachute deployment, heatshield separation, back-shell separation and retropropulsion firing. Each event must occur within a very narrow operational envelope and most must be triggered autonomously, based on estimates of the where the spacecraft is relative to the ground and how fast it is traveling.

Furthermore, each event must be executed flawlessly in the presence of potentially significant variability in Martian winds, atmospheric properties and surface topography. While several manned Mars mission architecture presentations exist, most conveniently gloss over the EDL details, thereby leading many people to conclude landing humans on Mars should be easy. The reality, as we shall see, could not be more different, which is why seasoned scientists and engineers tasked with designing the EDL architecture regularly use phrases such as 'Six Minutes of Terror' to describe the anxiety evoked by sending a manned vehicle to Mars. These engineers and scientists, many of whom have been involved in sending robotic missions to the Red Planet, know that landing a spacecraft on the surface of Mars represents the

Table 4.1 Generic entry maneuver timeline

Phase	Altitude (km)	Time to Touchdown Min:sec	Description
1. De-orbit burn.	400	31:05	Velocity 3.091 kms
2. Atmospheric re-entry begins	125	09:05	Commence deceleration.
3. Maximum deceleration.	70	08:00	1 G deceleration.
4. Thruster fires to turn vehicle to its entry orientation.	40	07:00	Vehicle travels 300km in 3 min.
5. Peak heating rate attained.	37	05:10	Velocity 6 to 7 km/s
6. Endure maximum G loading.	30	05:05	Approximately 15 to 20 Gs
7. Parachute deployed by mortar.	12	04:00	Velocity 1 kms
8. Heatshield jettisoned.	10	02:02	Flight path angle changes to vertical descent. Velocity 631 m/s.
9. Descent imager turned on.	1.5	01:00	Delta-V applied = 654 ms.
10. Heat shield jettisoned.	1.0 to 1.5	00:55	Velocity 2.4 m/sec. retro-propulsion ignited.
11. Vertical descent.	1.0 to 0	00:45	Earth Return Vehicle targeted. Adjustments made to bring vehicle to within 100m.
12. Landing.	0	00:00	Solar panels and antennae redeployed.

Table 4.2 Success and failure rate of Mars missions since 1975

Launch Date	Designation	Country	Result	Reason
1975	**Viking 1** Orbiter/Lander	US	**Success**	First successful landing on Mars.
1975	**Viking 2** Orbiter/Lander	US	**Success**	Returned 16,000 images.
1988	**Phobos 1** Orbiter	USSR	**Failure**	Lost en-route to Mars.
1988	**Phobos 2** Orbiter/Lander	USSR	**Failure**	Lost near Phobos.
1992	**Mars Observer**	US	**Failure**	Lost prior to Mars arrival.
1996	**Mars Global Surveyor**	US	**Success**	More images than all missions.
1996	**Mars Pathfinder**	US	**Success**	Performed several experiments
1998	**Nozomi**	Japan	**Failure**	No orbit insertion.
1998	**Mars Climate Orbiter**	US	**Failure**	Lost on arrival.
1999	**Mars Polar Lander**	US	**Failure**	Lost on arrival.
2001	**Mars Odyssey**	US	**Failure**	High resolution imaging of Mars.
2003	**Mars Express** Orbiter/Beagle 2 Lander	ESA	**Success/ Failure**	Orbiter imaging Mars. Lander lost on arrival.
2003	**Mars Exploration Rover** – Spirit	US	**Success**	Operating lifetime 15 times longer than expected.
2003	**Mars Exploration Rover** – Opportunity	US	**Success**	Operating lifetime 15 times longer than expected.
2005	**Mars Reconnaissance Orbiter**	US	**Success**	Returned more terabits of data than all other Mars missions combined.

most treacherous challenge of manned surface exploration, a reality reflected in the success and failure rate of robotic systems sent to Mars (Table 4.2).

The harsh reality of the EDL problem is that, to date, sixty percent of all Mars missions have failed and many of the successful ones have only been achieved by virtue of the spacecraft being small enough to allow them to reach the surface safely. Understandably, the problem of Mars EDL has generated much research, endless suggestions and volumes of articles but, to date, no one has solved the vexing problem of how a large manned vehicle traveling through the Martian atmosphere at Mach 5 can be decelerated to Mach 1 in ninety seconds and then re-orient itself from being a spacecraft to a lander, deploy parachutes and use landing engines to touch down. It's a quandary engineers often refer to as the 'Supersonic Transition Problem' but, before discussing this and some of the EDL challenges, it is useful to have an end-to-end overview of the key steps involved in the EDL process.

Generic entry, descent and landing sequence

Exoatmospheric flight

As the vehicle approaches the Martian atmosphere, EDL software will be initiated, venting of the internal heat rejection system will be conducted and, if necessary, final trajectory correction maneuvers will be performed. The onboard computer will then initialize attitude knowledge using a star scanner which, by checking the positions of the stars, is able to inform the vehicle and the crew exactly where it is relative to Mars. Once these systems checks have been conducted, the computer will command the cruise stage to separate from the vehicle. At this point the vehicle is just minutes from the interface with the Martian atmosphere. The onboard computer will continue to calculate and update its attitude and approach to the atmosphere, using entry reaction control system (RCS) thrusters to perform any last-minute course adjustments. Shortly before entering the atmosphere the autonomous guidance and navigation system will ensure the vehicle's attitude is aligned at the precise angle required to safely perform the EDL maneuvers.

Entry into Mars' atmosphere

Before descending to the surface, the vehicle must first reduce its velocity so it can enter into Mars orbit. Although such a maneuver can be performed propulsively, a cheaper method is to use an aerocapture maneuver, virtually eliminating the use of propellant to slow the vehicle.

An aerocapture maneuver uses the Martian atmosphere to alter the velocity of the vehicle without using any fuel, instead relying on the judicious use of aerodynamic forces. As the vehicle enters the Martian atmosphere, vehicle energy is dissipated through heat transfer, causing the vehicle to slow down. Once most of the spacecraft's velocity has been dissipated, the vehicle can then be configured to conduct the entry maneuver.

Entry maneuver

An entry maneuver involves the vehicle approaching the atmosphere from a hyperbolic trajectory, decelerating from high velocity and descending to the surface. The vehicle can be passive (ballistic) or actively controlled. If the vehicle follows a ballistic entry path, it will be guided prior to atmospheric entry before plummeting through the atmosphere as dictated by its shape. In contrast, an actively controlled vehicle may maneuver autonomously while in the atmosphere to confirm the best landing site. It is likely the entry maneuver will be combined with the deployment of some type of parachute to slow the vehicle to the velocity required for the final descent.

Parachute descent

Deployment of the supersonic parachute will occur when the vehicle is still traveling at up to Mach 2.0. The system will be deployed by a mortar, triggered when the inertial measurement unit (IMU) determines a fixed planet-relative velocity has been achieved. Shortly after the parachute is deployed, the heat-shield will be jettisoned, the descent stage exposed and two radar-based terminal descent sensors (TDSs) will be activated. During the parachute descent, the TDS will measure vehicle attitude and velocity relative to the Martian surface using an altimeter and Doppler velocimeter. Two TDSs are necessary due to the possibility that one of the TDS beams may be obscured by the separating heat-shield.

Powered descent
The beginning of the powered descent phase will occur while the vehicle is still descending by parachute and is initiated by the velocity and altimeter readings of the TDS. Before the powered descent phase begins, the lander engines will be warmed up for one or two seconds before the parachutes are jettisoned. Once the vehicle is clear of the parachute canopies, the engines will be throttled and the vertical velocity gradually decreased. Since the possibility of the parachute canopy or parachute system fouling the descent of the vehicle exists, it may be necessary to perform a divert maneuver during the descent profile.

Throughout the descent, hazard detection software and radar sensors will continue to operate, providing the onboard computer with retargeting information based altitude and horizontal velocity. The system will also generate new reference trajectories, depending on hazards detected and fuel expended.

Touchdown
Powered descent will continue until approximately one meter above the surface, resulting in a vertical velocity at touchdown of two to three meters per second. After four or more months in deep space and less than ten minutes after entering the Martian atmosphere, the vehicle will finally land on the surface of Mars and the crew will most likely breathe a collective sigh of relief!

On paper, this sequence of events appears relatively straightforward but, as described in the following section, the potential of the Martian environment to compromise one or more of these steps makes translating theory into reality a challenging and arduous task.

Why landing on Mars won't be easy

Atmospheric anomalies
Spacecraft returning to Earth have the luxury of entering a thick atmosphere capable of slowing a vehicle traveling at up to ten kilometers per second to less than Mach 1 while still twenty kilometers above the ground, just by using a heat shield. The rest of the return to Earth is achieved using drag, lift and parachutes. On Mars, however, the atmosphere is only one percent as dense as Earth's, meaning there is no air resistance to decelerate a vehicle. In fact, the atmosphere of Mars at its *thickest* is equivalent to Earth's atmosphere at approximately thirty-five kilometers altitude!

Upon arrival at Mars, the spacecraft will be traveling at four kilometers per second (Table 4.1) or faster. Due to the tenuous atmosphere, a heavy spacecraft, such as the one required for a manned mission, may never attain the subsonic terminal descent velocity required to commence the propulsive maneuvers for landing. The Martian atmosphere further compounds the EDL problem since, even if a vehicle finally manages to decelerate to subsonic velocity, the time for the remaining EDL events is very small, leaving little or no margin for error. In fact, by the time a vehicle is low enough to deploy subsonic decelerators, it may be too close to the surface to prepare for landing and, as some robotic Mars explorers have learned the hard way, it's not typically the fall that kills you, it's the landing!

Surface hazards

Using current technology, a manned Mars vehicle would initiate its landing system while still suspended on a parachute one kilometer above the surface. With only one thousand meters of altitude before touchdown, even the best landing system developed will be pressed for time to successfully complete the myriad hazard avoidance maneuvers expected during the terminal descent phase (TDP). The problem of detecting hazards is compounded by the fact that orbital surveillance is only able to detect rocks greater than one meter in size, although thermal inertia is capable of providing information concerning very rocky surfaces. As the vehicle descends at 2.4 meters per second, radar altimetry, Doppler radar and touchdown targeting algorithms will provide information regarding possible hazards, but these systems can be 'spoofed' by features such as slopes and other surface shapes. Furthermore, horizontal errors may be induced when a wide beam from a Doppler radar measures surface-relative velocity over slopes.

Non-redundant systems

NASA designs most of its most mission-critical systems with double or triple levels of redundancy, and with good reason, since failure of certain systems may result in loss of mission (LOM) and/or loss of crew (LOC). Due to the myriad ways LOM or LOC could occur during EDL, it is natural to assume EDL systems will be designed with multiple levels of redundancy, but this will not be the case. The reason is the very limited time span of the Mars EDL and the complexity of switching between systems in flight. As a result, most EDL subsystems must be designed to be non-redundant, a feature engineers refer to as *single-string*.

Landing accuracy

To land close to pre-positioned assets, the manned vehicle will need to be capable of exerting aerodynamic control over the atmospheric flight path, a feat yet to be realized in robotic missions. The vehicle will also need to be capable of autonomously adjusting its flight within the Martian atmosphere, requiring real-time hypersonic guidance algorithms which have yet to be written, although companies such as SEI are developing software to resolve this capability gap. Assuming the capability gap is resolved, the guidance algorithms will be designed to work in conjunction with the approach navigation software and inertial navigation system to help the vehicle perform a landing in close proximity to the pre-positioned assets. During the descent to the surface, the guidance algorithm, which will be automatically linked to the reaction control system (RCS), will maneuver the vehicle to accommodate off-nominal landing states and to take into account any unusually

Figure 4.1 Solving the entry, descent and landing problem has been likened to herding cats in a room filled with smoke and mirrors, where the floor is covered with apples and oranges. (NASA.)

atmospheric flight conditions such as surface winds, vertical gusts, and abnormally high or low atmospheric pressure.

Despite these known obstacles, there are few at NASA, or elsewhere, spending quality time attempting to resolve these issues. With its focus on retiring the Shuttle, qualifying the Ares I launch vehicle, completing the International Space Station (ISS) *and* returning humans to the Moon, the American space agency just doesn't have the resources to figure out how to make a reliable EDL architecture. Fortunately, there is much interest in the EDL conundrum throughout the space sector and solutions are being developed to tackle a problem that some engineers have described as analogous to herding cats in a room with smoke and mirrors, where the floor is covered with apples and oranges (Figure 4.1)!

RESOLVING THE EDL PROBLEM

The ability to develop, demonstrate and mature new EDL technologies will be a key to realizing the goal of landing humans on Mars but, in developing these technologies, engineers must not only contend with the problems posed by the atmosphere but also ensure the survival of a deconditioned crew. The transition from hypersonic to touchdown speed in a just a few minutes requires some ingenuity on the part of engineers to ensure deceleration forces do not disable the crew. Such a rapid transition from hypersonic to subsonic velocities also results in tremendous heat being generated, requiring adequate thermal protection strategies to be developed and, since EDL events occur so quickly, systems must also be designed to function with the highest system reliability of any spacecraft ever developed.

In the following section some of the more promising concepts are discussed, beginning with possible solutions to entry and concluding with an examination of techniques being developed to ensure a safe landing.

Approach and entry to Mars' atmosphere

Aerocapture
Entry into the Mars atmosphere begins one hundred and twenty-five kilometers above the surface with the spacecraft hurtling towards the Red Planet at more than sixteen thousand kilometers an hour. Obviously the first stage in ensuring a survivable EDL is to develop some means of safely reducing this speed, a challenge that may be achieved by utilizing a method known as aerocapture (Figure 4.2).

Figure 4.2 Aerocapture is a flight maneuver used to insert a spacecraft into orbit using the atmosphere as a brake. The atmosphere creates friction, used to slow the vehicle, transferring the energy generated by the vehicle's high speed into heat. The maneuver enables quick orbital capture without the requirement for heavy loads of propellant. (NASA.) See Plate 5 in color section.

In all Mars DRMs, the method of choice for capturing into Mars orbit is aerocapture, a fuel-saving technique whereby a spacecraft dives deep into the Martian atmosphere, gradually slowing itself down, before finally attaining an orbit around the planet. Because the vehicle uses the atmosphere as a means of shedding speed, aerocapture results in a significant mass reduction as less propellant is required. The method also offers mission planners the ability to accommodate uncertain atmospheric conditions and reduce the peak entry deceleration forces upon the crew, relative to a direct EDL in which a vehicle would head directly for the surface, without first attaining orbit. Unfortunately, these advantages are offset to a degree by the fact that aerocapture remains an untested technology!

Aerocapture challenges
Despite the advantages of aerocapture, the design of aerocapture trajectories is a challenge due to the several constraints that must be addressed by mission planners. For example, as the vehicle approaches Mars at hypersonic velocity, a trajectory must be calculated, resulting in an optimum flight path angle. If this angle is too shallow the vehicle will take a long time to decelerate, experience low heat load but will be subject to heat for a long period, a condition engineers refer to as a *high integrated heat load*. Conversely, if the angle is too high, the heat rate and deceleration will be very high but the vehicle will be subject to the heat for only a short period of time, a condition engineers refer to as a *low integrated heat load*. The challenge mission planners face is ensuring a flight path angle that does not subject astronauts to excessive g forces while at the same time ensuring the trajectory does not impose excessive thermal strain on the vehicle and crew. This trajectory is referred to by mission planners as the *entry corridor* and, as astronauts approach Mars, the guidance, navigation and inertial measurement systems onboard the vehicle will calculate the corridors' optimum width, hopefully ensuring a safe transit to a Mars parking orbit.

Evolved acceleration guidance logic for entry (EAGLE)
One way of ensuring the vehicle enters the optimum entry corridor may depend on special software such as the Evolved Acceleration Guidance Logic for Entry (EAGLE). EAGLE is a hypersonic guidance algorithm [11] being developed for manned Mars missions by scientists at the University of California and at NASA's Jet Propulsion Laboratory (JPL). EAGLE simply consists of a trajectory planner and a trajectory tracker. During the entry phase, the trajectory tracker sends attitude commands to the autopilot which calculates the correct angle of attack and the optimal aerodynamic lift force for the trajectory to be flown. Using information from accelerometers, aerodynamic acceleration is calculated and this information is sent to the trajectory tracker enabling EAGLE to update the reference trajectory accordingly and keep the vehicle within the entry/flight corridor. To test the effectiveness of EAGLE, the software has been evaluated using the Mars atmospheric flight simulation in the Dynamics Simulator for Entry, Descent and Surface Landing (DSENDS) spacecraft simulator at JPL.

DSENDS is a high-fidelity spacecraft simulator that helps mission planners

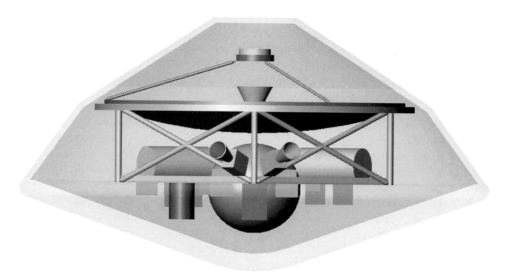

Figure 4.3 A rigid aeroshell encases the spacecraft, providing an aerodynamic surface and protection from the heat of entering the atmosphere at high speed. The aeroshell comprises an external thermal protection system, a supporting substructure surrounding the vehicle and adhesive bonding the thermal protection system to the substructure. (NASA.)

design EDL plans for vehicles landing on planets. In addition to simulating the actual flight of a vehicle through the Martian atmosphere, DSENDS is also able to integrate other computer models such as those for thrusters, star trackers, gyros and accelerometers. The end result is a simulator capable of providing mission planners with very accurate information about a specific EDL plan. Based on promising results at JPL, it would seem the problem of ensuring the correct flight path is on track to being resolved. The next step is to ensure the crew is protected during the stresses of deceleration and the thermal loads encountered as the vehicle passes through the atmosphere.

Aeroshells

As the vehicle passes through the upper regions of the Martian atmosphere, some deceleration will be achieved by the spacecraft itself since it will be fitted with a large diameter aeroshell, as much as fifteen meters wide (Figure 4.3).

One of the aspects of the blunt body design of the spacecraft is that in each flight phase acceleration is imparted upon the vehicle in the same direction. Hence, no vehicle reorientation is required during the EDL profile and there is no requirement to reorient the crew positions for tolerating *G* forces during the descent. While *G* forces may not present too much of a problem, the crew must also be protected from the thermal load on the vehicle. To ensure the crew is not subjected to excessive heat during entry, the thermal protection system (TPS), which forms a part of the aeroshell, must cope with the harsh aerocapture environment and the heat generated during the entry from orbit.

Perhaps the two greatest problems of designing such a rugged TPS include the structural concerns associated with thermal expansion, caused by extreme temperature cycles and the requirement to support orbital functionality without compromising the TPS. The latter problem presents engineers with a real predicament since, to ensure attitude control, communications and orbit-trim capabilities, the TPS must have openings for items such as thrusters, antennae and engines. Of course, one solution is to have two heat shields, one for aerocapture and one for entry from orbit, but this would result in a greater mass penalty and would also require a means to jettison the aerocapture heat shield. It is possible engineers may resolve these problems, but it is also possible an alternative concept such as an *inflatable* aeroshell may provide the answer.

Inflatable aeroshells

The hypersonic entry velocity might also be reduced by constructing very large inflatable aeroshells (Figure 4.4). Inflatable aeroshells provide a low-volume, low-

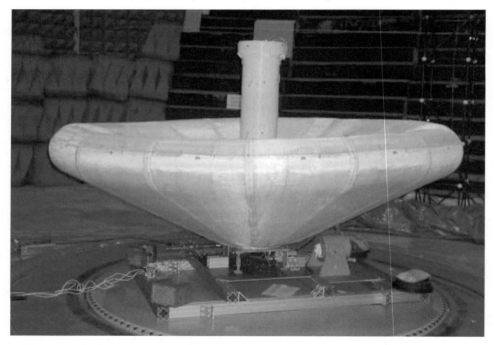

Figure 4.4 An inflatable aeroshell is characterized by a rigid nosepiece with an inflatable device attached around the perimeter. The nosepiece is covered by a thermal protection system to protect the structure from the heat generated during the aerocapture maneuver. Extending radially from the nosepiece is the inflatable aeroshell, which is covered by a flexible thermal protection system. The image shows a full-scale prototype unit built by ILC Dover. (ILC Dover/NASA.)

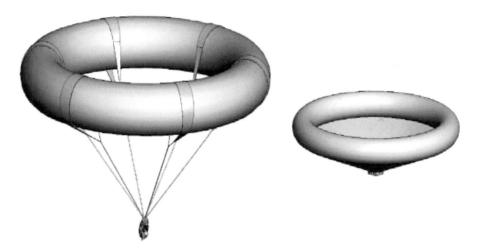

Figure 4.5 An artist's rendering of a Hypercone. (NASA.)

mass modular alternative to the rigid aeroshells discussed in the previous section. More importantly, the very nature of an inflatable aeroshell permits larger sizes to be deployed than are realistically achievable with rigid aeroshells. Furthermore, deploying such large aeroshells will result in a more benign thermal environment for the crew during entry. Although much work will be required to qualify inflatable and deployable aeroshells for a manned Mars mission, the technology is robust and NASA has already progressed from computer modeling studies to planning flight tests at Wallops Flight Facility (WFF) in conjunction with Langley Research Center (LRC). While the inflatable aeroshell concept appears promising in the near-term, some engineers are studying more advanced concepts that may prove even more effective. One such idea is the Hypercone.

Hypercone

The Hypercone (Figure 4.5) is an inflatable supersonic decelerator which, although only existing as a computer generated image (CGI), may prove to be the solution to slowing the speeding spacecraft faster than any other concept.

The donut-shaped Hypercone would be thirty to forty meters in diameter, would girdle the vehicle and be inflated with gas rockets, creating a conical shape. Inflation would occur at an altitude of ten kilometers while the vehicle is traveling at Mach 4 or 5. Acting as an aerodynamic anchor, the Hypercone would decelerate the vehicle to Mach 1, a velocity enabling the deployment of subsonic parachutes. Although the Hypercone could be used to decelerate a vehicle during atmospheric re-entry, it is only intended to supplement other deceleration mechanisms, bridging the gap in capability between conventional heat shields and conventional parachutes, which obviously wouldn't work in the Martian atmosphere.

Research is already underway at Vertigo Inc., in Lake Elsinore, California, where scientists are studying and evaluating possible sizing, mass estimates and materials for Hypercones designed to decelerate human-rated Mars vehicles. While Vertigo has decided the best material for the Hypercone will probably be a matrix of fabrics such as silicon and Vectran, the next phase of research requires supersonic wind tunnel testing, which is why the company has approached NASA for funding. Meanwhile, the *ballute*, a close relative to the Hypercone, is yet another deceleration concept being investigated

Ballutes

A deceleration solution similar to the Hypercone is the ultra lightweight ballute (ULWB), an innovative concept involving the deployment, prior to atmospheric entry, of a large, lightweight, inflatable aerodynamic decelerator.

The large drag area of the ballute enables the vehicle to decelerate even in a Martian atmosphere and, thanks to the lightweight construction of the system, the ballute allows more payload to be carried by the vehicle. As the name suggests, a ballute is a cross between a balloon and a parachute. During ballute aerocapture, the vehicle approaches Mars from interplanetary space and deploys the ballute, before entering the atmosphere. Once the vehicle enters the atmosphere, the vehicle begins to decelerate and, once the desired velocity is achieved, the ballute is released, enabling the vehicle to exit the atmosphere, where it can raise periapsis and achieve the desired orbit.

ULWB technology is currently being developed using funding provided by NASA's In-Space Propulsion (ISP) Office [9, 10, 13] and research has already been conducted to investigate the feasibility of ballute aerocapture [5, 6, 12]. The result of this research and development is what may be referred to as a Reference Ballute Configuration (RBC).

The four subcomponents of the RBC include the inflatable torus, a thermal protection system, inflatable booms and local reinforcement. The ballute comprises a multi-component laminate, composed of materials that are very low weight and capable of retaining their strength at temperatures as high as eight hundred degrees Celsius. The strength and performance characteristics of the ballute were initially tested using computational fluid dynamics (CFD) models, which computed drag efficiency, aerodynamic loads, aeroheating and flow stability for various ballute configurations. Once the CFD models were validated, hypersonic testing was conducted at NASA's Langley wind tunnel, where models were subjected to Mach 10 testing. This testing and evaluation allowed researchers to determine the optimum trajectory of a ballute entering the atmosphere and also enabled validation of the CFD data.

In addition to providing mass performance advantages, ballutes offer similar advantages to the Hypercone such as low heating rates during atmospheric entry and component packaging not constrained by the aeroshell structural envelope. Before the system is validated for operational mission applications it will require flight demonstration tests but the results to date indicate the ballute concept has the potential to fly lighter and fly cooler.

Stronger and larger parachutes

The promising advances in aeroshell and aerocapture techniques described will hopefully reduce the velocity of the spacecraft to Mach 3 or Mach 2. Further reductions in the velocity will require the use of some very strong supersonic parachutes. These parachutes will need to be constructed from heat-resistant fabrics capable of resisting extremely high inflation pressures. However, low density, high Mach tests here on Earth have demonstrated dynamic stability at speeds as high as Mach 2.5 [14].

Once a super-strong parachute fabric has been tested, it would be possible to increase the diameters of the parachutes, thereby increasing the deceleration capability of the parachute. Although this sounds feasible in theory, the reality may be a little more challenging since computer modeling has shown that to decelerate a fifty-tonne vehicle from Mach 3 to the fifty-meters-per-second velocity required near the Martian surface would require a supersonic parachute ninety meters in diameter [1]! Such a large parachute exceeds by several orders of magnitude the experience with parachutes utilized in the robotic exploration of Mars, which have generally been ten to twenty meters in diameter and deployed at speeds below Mach 2.1 [1]. A partial solution to slowing a vehicle using supersonic parachutes could be clustered supersonic parachutes, but the large timeline penalties for deploying the canopies would still present an obstacle. While advances in supersonic parachutes continue, some engineers searching for a deceleration solution are investigating supersonic retropropulsion.

Supersonic retropropulsive systems

A supersonic retropropulsion system would work by placing the thrusters' retronozzles on the forebody of the vehicle and firing directly into the direction of travel. Although there is little practical experience firing thrusters into a high dynamic pressure supersonic flow, modeling the effects of such a system is currently being addressed by engineers and scientists working with advanced CFD software.

Results of the CFD studies have shown that interaction of the jets with the opposing supersonic stream may cause an unstable flowfield around the vehicle. This instability is mainly due to the bow shock occurring as the vehicle tries to displace the supersonic stream. This instability may be low at high Mach numbers and high at low Mach numbers and may also be affected by the configuration of the retronozzles and the effect of firing the jets. Each of these factors has an effect on the stability of the vehicle and the problem faced by engineers out how to ensure stability through all the Mach regimes, while still allowing the jets to be fired. This conundrum is often referred to by engineers as a *stability transition phenomenon* (STP) and the problem is still not thoroughly understood. In order to understand how the STP can be resolved, scientists have experimented with various retropropulsion configurations to examine how the bow shock is affected at various supersonic speeds, using CFD modeling.

One arrangement examined is the central retropropulsion configuration (CRC), but several studies have shown a centrally-located retronozzle design is unfavorable from an aerodynamic standpoint [3, 7] due to the unsteadiness of the supersonic flow during most Mach regimes. Another arrangement tested has been the peripheral retropropulsion configuration (PRC), a more promising design where the retro-nozzles are located at the periphery of the forebody. Such a configuration, with three nozzles at the body periphery, has been tested at various angles of attack, different Mach numbers, and varying thrust forces. The test results revealed few pressure perturbations at the edges and confirmed the aeroshell surface was covered with a nearly uniform region of high pressure. Furthermore, the jet from the nozzle was able to penetrate the bow shock, suggesting the PRC may have potential as an optimum retropropulsion design.

Other studies have examined the dynamics of throttling effects using the PRC. As the vehicle enters the supersonic freestream of the Martian atmosphere, it will probably be necessary to alter pitch to avoid excessive heating and, also, to ensure the vehicle is aligned with the entry corridor described earlier. In the PRC design, the throttling on one engine and throttling combinations of two or more engines will obviously affect drag and may induce pitching moments, causing the vehicle to be unstable. Unfortunately, the research conducted on PRC throttling is limited but past experimental work has demonstrated supersonic retropropulsion to be possible, albeit on a small scale, and research continues to investigate propulsive-aerodynamic interactions of this concept.

A recent study [8] investigated supersonic retropropulsion, using a capsule similar to Orion, which will serve as the crew vehicle for NASA's Mars mission. Using aerodynamic modeling, the study examined various vehicle masses and aeroshell diameters during the retropropulsion phases under varying atmospheric conditions. The study team was particularly interested in the significance of aero-propulsive interactions that might occur at different altitudes and Mach speeds and, also, what effect initiating retropropulsion earlier or later in the trajectory might have on the ability of the system to decelerate the vehicle. The results were promising, demonstrating supersonic retropropulsion to be a technology potentially capable of achieving the landed masses for human Mars exploration.

The Skycrane option
Assuming NASA's Mars Science Laboratory (MSL) lands safely on the surface of Mars in 2009, the agency will no doubt be grateful to Sikorsky and the design of its S-64 Skycrane heavy-lift helicopter. NASA's version of the Skycrane describes a means of landing the MSL on the surface of Mars using retrorockets rather than rotor blades to lower its payload. The landing concept, in common with the Sikorsky, utilizes a winch and tether. Once the MSL is on the surface, the tether connection will be severed and the Skycrane will fly a short distance and crash land.

Concept of operations
Following a guided entry phase, a supersonic parachute is inflated to slow the vehicle down to speeds at which the terminal descent phase can begin. At this point, the vehicle comprises a Descent Stage (DS) and the MSL rover, rigidly attached and known collectively as the Powered Descent Vehicle (PDV). Located on the DS are

eight throttleable lander engines. During the parachute phase the heat shield is jettisoned, enabling the Terminal Descent Sensor (TDS) to provide altitude and velocity data. Pyrotechnics then separate the PDV from the parachute, and the lander engines go through their warm-up cycle at about two-and-a-half-kilometers above the surface. At two kilometers above the surface, the terminal descent phase (TDP) begins. During the TDP, the MSL rover separates from the PDV but remains attached by a seven-and-a-half-meter bridle. Once touchdown of the rover is confirmed, the bridle is cut and the DS performs a controlled fly-away and crash lands a few hundred meters away from the MSL rover

While the Skycrane concept may work for the 775 kilogram MSL, adapting the idea to work for thirty-tonne payloads may require innovative engineering. However, many in the space industry consider the Skycrane to be the most elegant and simple system devised to place payloads on the surface of Mars and it is possible the concept can be scaled for larger manned vehicles.

Space elevator
A less dynamic solution than the Skycrane is the space elevator. Although such a concept may seem far-fetched, space elevators may be closer to reality than many people imagine. The NASA Institute for Advanced Concepts (NIAC) recently commissioned Dr. Bradley C. Edwards to study all the aspects for constructing and operation of a space elevator. The report [4] describes in detail all the facets of designing and operating such a system.

To begin with, the elevator would not be a cable but would start life as a very thin piece of tape tens of thousands of kilometers long, tapering just five centimeters wide on the Earth's surface to about twelve centimeters wide in the middle. The tape, along with booster rockets, would be taken into orbit by a launch vehicle. The tape would then be flown back down to the Earth's surface while the booster rockets provided the necessary counterbalance in orbit (Figure 4.6).

Centripetal force ensures the higher part of the tape is thrown away from the planet, while the effect of gravity on the lower mass of the tape maintains tension. This first tape represents a baseline tape permitting 'lifter's to add supplementary tapes to increase the strength of the elevator to a useful capability. On completion, the space elevator would be capable of carrying several tonnes to orbit and spacecraft would simply ride up and down on an electrically powered climber, fuelled by ground-based lasers shining onto solar panels.

Constructing such a device will require a very strong material but fortunately, such a material has existed for almost two decades. Carbon nanotubes, first produced in 1991, are nanostructures which are the strongest and stiffest materials on Earth in terms of tensile strength. In fact, Carbon nanotubes are so small that fifty thousand of them would fit inside a human hair. The only problem remaining to be resolved is how to produce tapes thousands of kilometers long, since the present capability is only a few centimeters.

Assuming the production problem is resolved, the next challenge is powering the elevator. In Arthur C. Clarke's "The Fountains of Paradise" [2] a space elevator, which is used to enable people settle in orbit, is powered by nuclear fusion and

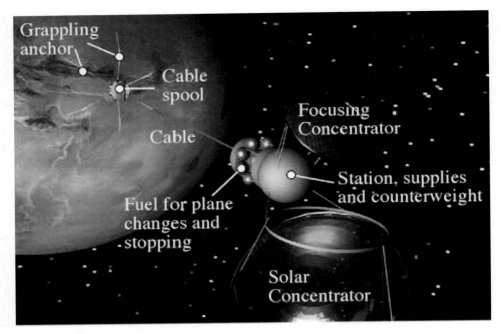

Figure 4.6 An artist's rendering of a space elevator. (NASA.)

superconductors, but the NIAC report suggests a less exotic power system using solar arrays and batteries.

Once the production and power aspects of the elevator have been resolved, the system would need to be proven on Earth before being used on Mars. Unfortunately, this may take some time, since it is predicted that the construction of an Earth-based space elevator would take ten years and cost as much as $40 billion, which also happens to be how much the cheapest manned Mars mission is projected to cost! So, even if the production issues are solved in the next ten years and the approval is given to construct a space elevator, it is unlikely such a system could be used to transport astronauts from Mars orbit to the Martian surface anytime soon. Despite the daunting problems facing the nanotube community, space elevator enthusiasts remain optimistic, insisting one day, astronauts will depart from Martian orbit to the surface without using a drop of rocket fuel and without deploying a single parachute.

IN SUMMARY

Although the United States has successfully landed six robotic systems on the surface of Mars, it is evident that significant technology and engineering investment will be required to achieve the EDL capabilities required to safely land humans on the surface of the Red Planet. However, as robotic exploration technology evolves, improvements will be made to current EDL system delivery until, eventually, by improving the systems described in this chapter, or by combining one or more of the technologies, it will be possible to safely land a large mass system associated with human Mars exploration.

REFERENCES

1. Braun, R.D., and Manning, R.M. Mars Exploration, Entry, Descent and Landing Challenges. IEEEAC paper #0076. December 9, 2005.
2. Clarke, A.C. Fountains of Paradise. Aspect. September, 2001.
3. Daso, E.O.; Pritchett, V.E., and Wang, T.S. The Dynamics of Shock Dispersion and Interactions in Supersonic Freestreams with Counterflowing Jets. 45th AIAA Aerospace Sciences Meeting, AIAA 2007-1423, Reno, Nevada, January 2007.
4. Edwards, B.C. The Space Elevator. NIAC Phase II Final Report, March 1, 2003.
5. Gnoffo, P.A., and Anderson, B.P. Computational Analysis of Towed Ballute Interactions. AIAA 2002-2997, 8th AIAA/ASME Joint Thermophysics and Heat Transfer Conference, St. Louis, MO, June 24–26, 2002.
6. Hall, J.L., and Le, A.K. Aerocapture Trajectories for Spacecraft with Large, Towed Ballutes. 11th Annual AAS/AIAA Space Flight Mechanics Meeting, Santa Barbara, CA, AAS 01-235, February 11–15, 2001.
7. Jarvinen, P., and Adams, R. The Effects of Retrorockets on the Aerodynamic Characteristics of Conical Aeroshell Planetary Entry Vehicles. AIAA 70-219, AIAA 8th Aerospace Sciences Meeting, New York, New York, January 1970.
8. Korzun, A.M. Supersonic Retropropulsion Technology for Application to High Mass mars Entry, Descent and Landing. Georgia Institute of Technology. April 2008.
9. Masciarelli, J.; Miller, K. Ultralightweight Ballute Technology Advances. Proceedings of the 2nd International Planetary Probe Workshop, NASA Ames Research Center, Moffett Field, CA, August, 2004.
10. Masciarelli, J.; Miller, K.; Rohrschneider, R.; Morales, A.; Stein, J.; Ware, J.; Lawless, D.; Westhelle, C.; Gnoffo, P.; Buck, G.; McDaniel, J. Fly Higher, Fly Lighter, Fly Cooler: Progress in Ultralightweight Ballute technology Development. 53rd Joint Army-Navy-NASA-Air Force (JANNAF) Propulsion Meeting, Monterey, CA, December, 2005.
11. Mease, K.D.; Leavitt, J.A.; Benito, J.; Talole, A.; Sohl, G.; Ivanov, M.; Ling, L. Advanced Hypersonic Entry Guidance for Mars Pinpoint Landing. NSTC-07-0023. NASA Conference. 2007.
12. Medlock, K.L., and Longuski, J.M. An Approach to Sizing a Dual-Use Ballute System for Aerocapture, Descent and Landing. 4th International Planetary Probe Workshop, Pasadena, CA, June 27–30, 2006.
13. Miller, K.L.; Gulick, D.; Lewis, J.; Trochman, B.; Stein, J.; Lyons, D.; Wilmoth, D. Trailing Ballute Aerocapture Concept and Feasibility Assessment. AIAA 2003-4655, 39th AIAA/ASME/SAE/ASEE Joint Propulsion Conference, Huntsville, AL, July, 2003.
14. Witkowski, A., and Brown, G. Mars Deployable Decelerators Capability Roadmap Summary. 2006 IEEE Aerospace Conference, Paper 1585, Big Sky, Montana, March, 2006.

5

Propulsion systems

"Any sufficiently advanced technology is indistinguishable from magic."

Arthur C. Clarke

Although much research is being conducted on advanced propulsion technologies capable of supporting human interplanetary travel, much of this research is focused upon concepts that, while offering significant performance improvements over present systems, have fundamental scientific issues that must be resolved before they can be seriously considered. For example, fusion and antimatter are both appealing as a propulsion option for manned Mars missions but, in the case of fusion, scientists are still a long way from demonstrating a device that has energy gains sufficient for commercial power. Equally, antimatter's high density makes it attractive to mission designers but the high costs of production make its use impractical. Although many of the issues facing these exotic propulsion options will eventually be overcome, it is unlikely these technologies will be ready for space applications by the time the first manned mission to Mars is ready to embark. For mission planners struggling with ways of protecting astronauts from deep space radiation for four months or more, this state-of-affairs is disappointing to say the least, but in the present fiscal environment it would require a prohibitive investment to resolve the issues related to the sophisticated technologies mentioned.

To achieve a quantum increase in propulsion capability by the time a manned Mars mission is ready to embark, it will be necessary to pursue safe, affordable systems with very high power densities that should be based on the physics of today. Based on these constraints, the requirement for high power density eliminates all propulsion methods except nuclear energy sources and the emphasis on known physics and affordability reduces the scope to fission processes such as nuclear thermal, gas-core and internal and external nuclear pulse, all of which are politically unacceptable. However, the chemical rocket alternative is unacceptable from an astronauts and mission planner perspective for reasons of safety, since chemical rocket methods result in Earth-Mars transit times of four months or more. Such long transit times expose crewmembers to deep space radiation for long periods of time and result in pronounced bone deconditioning. Furthermore, chemical rocket methods offer few abort options to the crew in the event of a contingency.

Obviously, to ensure the safety of crew, an interplanetary vehicle must not only be fast and reliable but must also provide abort options in the event of a systems malfunction or crew contingency.

This chapter describes some the more viable propulsion options that may be available to mission planners during the next two to three decades and also briefly discusses some of the more exotic technologies that may be available for later missions.

VARIABLE SPECIFIC IMPULSE MAGNETOPLASMA ROCKET

Overview

The Variable Specific Impulse Magnetoplasma Rocket (VASIMR^{TM}) was designed by seven-time Space Shuttle astronaut, Franklin Chang-Diaz (Figure 5.1), who has been working on the idea since 1979. A high-power, plasma-based propulsion technology, the VASIMR^{TM} is being privately developed by the Ad Astra Rocket Company, based in Webster, Texas, just three kilometers from NASA's Johnson Space Center (JSC). Ad Astra was established in January 2005 to commercialize the technology of the VASIMR^{TM} and as this book is being written the company is preparing to conduct an initial test firing of a full-scale engine and also to fly a scaled down version onboard the International Space Station (ISS).

Figure 5.1 US astronaut Franklin Chang-Diaz working on the International Space Station during Space Shuttle mission STS-111. (NASA.) See Plate 6 in color section.

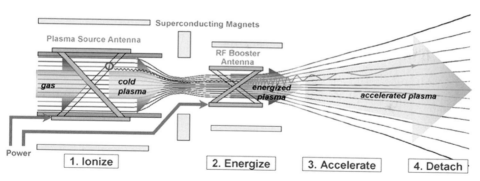

Figure 5.2 A simplified diagram of the Variable Specific Impulse Magnetoplasma Rocket. The Plasma Cell is where hydrogen is injected to be turned into plasma. The RF Booster amplifies and energizes the plasma to the required temperature and the Magnetic Nozzle converts the energy of the plasma into thrust. (Ad Astra.)

Plasma-based propulsion technology
In developing the VASIMR^{TM}, the goal of Chang-Diaz and his research team was to achieve high fuel efficiency. Since high fuel efficiency couldn't be achieved using conventional chemical technology, the team focused instead on plasma physics, in which temperatures of $10,000^{0}C$ strip atoms of their electrons, resulting in a mix of positive electrons and negative ions. At these extremely high temperatures, the ions, which comprise the majority of the mass, move at velocities sixty times faster than particles in even the most efficient chemical rocket. To achieve such high temperatures, the VASIMR^{TM} [2, 3] employs electromagnetic waves similar to the ones utilized in a microwave oven.

To achieve the propulsive force, three events must occur inside the rocket engine (Figure 5.2). First, hydrogen must be injected at the forward-end cell of the engine and ionized. Once electrically charged, the hydrogen is then heated by electromagnetic waves to produce a specific density in the central cell of the engine. Once the hydrogen is heated to a very high temperature, it becomes plasma and enters a two-stage hybrid nozzle at the aft end of the rocket, at which point it is exhausted to produce thrust [2, 3].

One of the advantages of the VASIMR^{TM} propulsion technology is the capability of constant power throttling (CPT), which is a function akin to the transmission in a car when climbing a hill or feathering a propeller engine while flying an aircraft. This means a spacecraft utilizing this type of propulsion technology will be capable of using its propulsion to enter into Mars orbit, rather than relying on many of the systems described in the previous chapter.

The rocket engine
The VASIMR^{TM} rocket (Figure 5.3) comprises a complex system of superconducting magnets, tightly packaged power generation and conditioning mechanisms, robust radio frequency systems, a hybrid magnetic nozzle and lightweight heat

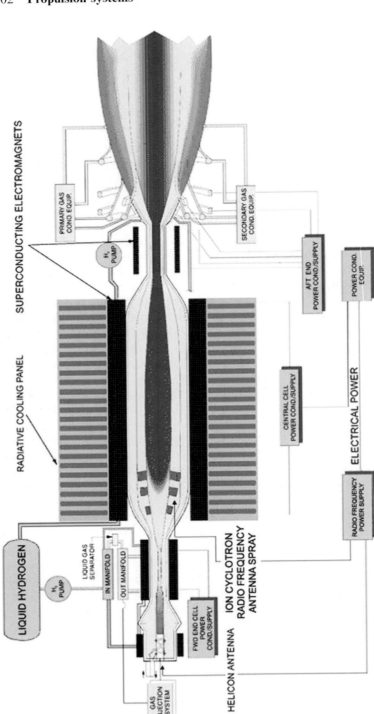

Figure 5.3 Components of the Variable Specific Impulse Magnetoplasma Rocket. (NASA.)

shields [2, 3]. The 200-kilowatt VX200 VASIMR™ system being readied for flight tests is the product of twenty-five years of an expanding research effort involving a talent pool of some fifty scientists drawn from seven universities and two national laboratories.

VASIMR today

In common with many visionary propulsion projects, the VASIMR™ has struggled to obtain funding. In fact, as Chang-Diaz sat in his seat aboard the Space Shuttle Endeavour waiting for the launch of STS-111 and his record-breaking seventh Space Shuttle mission, NASA project managers were actually discussing closing the project!

Part of the reason VASIMR™ has struggled for funding is the disagreement among space propulsion experts regarding the flaws and benefits of this revolutionary technology. Some propulsion specialists claim that for a VASIMR™ rocket to be capable of transporting humans to Mars it would be necessary to boost the power to more than ten megawatts. This amount of power, critics argue, could only be achieved by using a very large onboard nuclear reactor, which will not exist for some time. VASIMR™ supporters argue the concept serves as a precursor to the Holy Grail of rocket propulsion: a fusion rocket. Although controlled fusion has yet to be realized, the efforts to achieve it have been relentless and the progress encouraging, as evidenced by the efforts of Chang-Diaz and his team. Meanwhile, the VASIMR™ continues to be on trial, although the successful test-firing of the VX200 in 2008 will likely mean more cash directed to Ad Astra, with the goal of testing in space. At the time of writing, the company is in negotiations with the United States government to mount a prototype rocket engine on the ISS for testing. The flight to the ISS could occur in 2011 or 2012 and, since the Space Shuttle is due to retire in 2010, the engine would probably be launched onboard a commercial launch service. Once delivered to the ISS, the engine would be installed during an EVA and would be operated either from inside the station or remotely from ground stations. Once the engine has proven its capabilities, it could be used to reposition the ISS.

Although VASIMR™ remains a work in progress, it is an expanding research effort developing something that is very new and very different from the established chemical propulsion framework. More importantly, the system possesses attributes that are very attractive to both astronauts and engineers since a twelve megawatt version could deliver twenty metric tonnes to Mars in less than one hundred and twenty days, while a two hundred megawatt version could achieve the trip in just thirty-nine days! People have touched on the Elephant in the Room issue of the public's fear of anything nuclear, a concern exaggerated by the media, but the fact is the United States Navy has operated nuclear power plants in the eight megawatt range for decades without a single incident that could have led to a core failure. The VASIMR™ engine is clearly a revolutionary technology that may well be the deal breaker enabling fast missions to and from Mars and, eventually, to realizing a long overdue revolution in spaceflight propulsion.

NUCLEAR PROPULSION

History

Nuclear propulsion has never been a popular option among the general public thanks largely to misleading and erroneous information generated and perpetuated by a misinformed media. For example, when the Cassini probe destined for Saturn was launched in 1997, its thirty-three kilograms of plutonium caused an uproar, resulting in more than a thousand protesters bringing their message to the gates of Cape Canaveral Air Force Station. The protesters mistakenly believed thousands of people would be exposed to lethal doses of radiation if something went wrong, ignoring the fact the Cassini mission marked the twenty-fourth time generators containing plutonium had been launched and, even if the rocket exploded on the launch pad, the canisters containing the plutonium were designed to withstand such an accident. Furthermore, even if the canisters were compromised, the increase in radiation resulting from the destruction of Cassini would have amounted to only one fifteen thousandth of a normal lifetime's absorption of radioactivity, or about the same as a dental X-ray!

The point is that safety can be assured and the benefits are great. So, why is NASA not building nuclear powered spacecraft to take us to Mars in weeks, rather than months? The reason is not technical, but political.

Nuclear thermal propulsion overview

One of the leading propulsion options for future manned Mars missions is nuclear thermal propulsion (NTP) in which hydrogen is heated to extreme temperatures and ejected at high velocities, as in a conventional rocket. Such a design would enable very fast changes in velocities and would reduce the time taken to travel to Mars from several months to just a few weeks or even days. Another advantage of NTP is that any gas maybe used for propulsion, although some gases are better than others, a feature enabling an NTP spacecraft to use indigenous Martian atmosphere or even water ice to refuel, thereby prolonging its lifetime and reducing cost.

The NTP method promised to be so effective that NASA began its Nuclear Engine for Rocket Vehicle Application (NERVA) program in 1963 to develop a thermal nuclear propulsion system for use on interplanetary missions. One of the nuclear rocket engines built as a part of the NERVA program was the Phoebus-2A which delivered an I_{sp} of 825 seconds. Despite the promise of such a powerful rocket engine, the NERVA program was cancelled in 1973 as a result of environmental concerns and lack of public and political support for a manned mission to Mars.

Nuclear thermal propulsion technology

A NTP creates thrust by heating and expanding a fusion fuel, such as hydrogen, in a nuclear reactor. The high pressure fuel flows through pumps, cooling the rocket engine nozzle before continuing through the reactor pressure vessel, neutron reflector, control drums, core support structure and internal radiation shield. During the process, the fuel collects heat to drive the turbines, and the hydrogen exhaust is then channeled through coolant channels in the reactor core's fuel elements. Here, the hydrogen absorbs energy released by fissioning uranium atoms and is superheated before being expanded out of a supersonic nozzle for thrust, controlled by matching the turbopump supplied hydrogen flow to the reactor power level. Controlling the reactor power level is achieved by control drums regulating the number of fission-released neutrons reflected back into the core. To prevent radiation escaping, an internal neutron and gamma radiation shield containing interior coolant passages is located between the reactor core and sensitive engine components.

Nuclear thermal propulsion today
NTP concepts have already been tested by NASA as a part of design studies funded by NASA's Nuclear Propulsion Office (NPO) between 1992 and 1993, which produced a small NTP rocket producing an I_{sp} of more than nine hundred seconds [4]. A derivative of the nuclear thermal rocket (NTR) was also configured to provide both propulsive thrust and power generation for crew life support. Referred to as bimodal NTR (BNTR), such a system is possible because only a relatively small amount of the enriched uranium-235 fuel is actually consumed during primary propulsive maneuvers [1]. As a part of the studies, a concept BNTR transfer vehicle was designed to demonstrate how few transportation system elements would be required and also how artificial gravity could be integrated into the BNTR vehicle.

Bimodal nuclear thermal rocket (BNTR)
The concept BNTR vehicle [1] includes a bimodal core stage connected to an inflatable TransHab via a saddle truss that is open underneath to allow the liquid hydrogen tank to be jettisoned following trans-Mars injection (TMI). Once the tank is released, the crew transfer vehicle initiates vehicle rotation at four revolutions per minute, providing the crew with a Mars gravity environment during the outbound leg. The BNTR transfer vehicle (Figure 5.4) would utilize a common core stage outfitted with three BNTR engines capable of providing up to fifty kilowatts of electrical power using any two engines. Unlike many of the mission architectures described in Chapter 3, the bimodal core stage would not be jettisoned after performing the trans-Mars injection (TMI). Instead, it would remain attached to the rest of the vehicle to provide midcourse propulsive maneuvers, necessary power during the Earth-Mars transit and to perform the aerobraking maneuver placing the vehicle in low Mars orbit (LMO). This particular feature reduces the operational complexity of the mission by not only eliminating the need for an aerobrake and injection stage but also removing the requirement for solar array deployment and retraction [1].

In common with Chang-Diaz's VASIMR^{TM} concept, BNTR represents an evolutionary approach for developing new space propulsion systems and, assuming the concept can be developed affordably, has tremendous potential to enhance or enable deep space missions. However, the liquid oxygen (LOX)-augmented NTR, a variation of the BNTR, may offer even greater mission performance benefits.

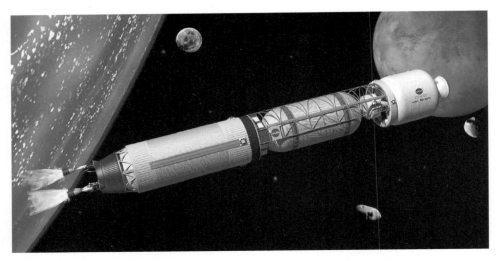

Figure 5.4 Artist's rendering of Bimodal Nuclear Thermal Rocket. (John Frassanito and Associates/NASA.) See Plate 7 in color section.

LIQUID OXYGEN AUGMENTED NUCLEAR THERMAL REACTOR

Technology
The enhanced NTR variant known as the LOX-augmented NTR (LANTR), combines the conventional liquid hydrogen-cooled NTR concept with supersonic combustion ramjet technology. LANTR [1] utilizes the large divergent section of the NTR nozzle as an afterburner into which LOX is injected and supersonically combusted with reactor-preheated hydrogen, eventually emerging from LANTR's sonic throat. By adjusting the oxygen-to-hydrogen mixture ratio, the LANTR engine could operate over a range of thrust and specific impulse levels while the reactor core power level remains relatively constant.

The LANTR system is versatile not only because it can operate over a range of thrust values but also because it can work as a conventional liquid hydrogen cooled NTR, a bipropellant LOX/LH2 engine or a power reactor. Furthermore, the high thrust generated by the LANTR engine provides the crew with a higher safety factor, compared with conventional propulsion, when entering Mars orbit. The LANTR transfer vehicle, in common with the VASIMR and BNTR, is able to propulsively capture into orbit rather than relying on an awkward and risky combination of ballutes, parachutes and supersonic retropropulsion techniques [1]. Finally, in addition to its versatility, the LANTR engine makes it possible to obtain big engine performance from a comparatively small and inexpensive nuclear engine, a feature attractive to mission planners who must figure out how to launch all the mission elements into low Earth orbit (LEO).

More than three decades have gone by since NASA's NERVA program was shut down but, with recent advances in technology, nuclear propulsion and its variants are no longer considered unrealistic or dangerous. In fact, if Mars is to be explored at reasonable cost and within reasonable mission times, space nuclear propulsion may be a requisite of interplanetary space missions. Of the various nuclear options a bimodal NTR or LANTR system clearly provides mission planners with the most versatile and safest interplanetary space transportation option.

MAGNETOPLASMADYNAMIC THRUSTERS

Overview
The magnetoplasmadynamic (MPD) thruster, a variation of the VASIMR, is the most powerful form of electromagnetic propulsion, capable of efficiently converting megawatts of electric power into thrust. MPD offers significant advantages over conventional types of propulsion due to the high exhaust velocities generated as a result of interaction between a magnetic field and an electrical current, a process known as *ion propulsion*.

Magnetoplasmadynamic thruster technology
The key to MPD is the ion, an atom or molecule that is electrically charged. The process of electrically charging an atom or molecule is *ionization*, in which electrons are added or removed, a process resulting in ions being either negatively (when ions gain one or more electrons) or positively (when ions lose one or more electrons) charged. Once all the atoms or molecules contained in a gas are converted into ions, a gas is considered to be ionized. In the case of MPD, plasma is an electrically charged gas in which all the negative and positive charges add up to zero. Possessing some of the properties of a gas, plasma is affected by both electric and magnetic fields and is a good conductor of electricity, making it ideal for use in electric propulsion.

A basic MPD thruster comprises a central rod-shaped cathode and a cylindrical anode surrounding the cathode. A high-current electric arc is generated between the anode and the cathode and as the cathode heats up, it emits electrons which collide with and ionize a propellant gas to create plasma. The process creates a self-induced magnetic field similar to when an electrical current is passed through a wire. The magnetic field interacts with the electric current passing through the plasma from the anode to the cathode, resulting in an electromagnetic force pushing the plasma out of the engine, thereby creating thrust.

Magnetoplasmadynamic research today
MPD has been sporadically researched over the years with much of the research for space applications being conducted at NASA's Jet Propulsion Laboratory (JPL) and Glenn Research Center (GRC). Current research at GRC is aimed at developing high-specific impulse, megawatt-class thruster technology that may one day be used to transport humans to Mars. At power levels of one megawatt, exhaust velocities of 100,000 meters per second have been demonstrated, a speed equivalent to eleven times the redline speed of the Space Shuttle! Unfortunately, these exhaust velocities

have been demonstrated for only short periods of time and it may be some time before a MPD thruster is developed that is capable of operating for the several weeks required for a Mars mission.

MAGNETIZED TARGET FUSION

Overview
Magnetized Target Fusion (MTF) is a relatively new propulsion concept with the potential for low-cost development and to utilize existing facilities and technology. The features making MTF attractive to spacecraft engineers and mission designers include the low system mass and volume, high thrust, and efficient means of operation, resulting in high gain and low waste.

Magnetized target fusion technology
The MTF concept (Figure 5.5) generates thrust by first creating target plasma through magnetic forces. The target plasma is then imploded by a spherical plasma liner formed from the merging of several plasma jets triggered by a set of plasma accelerators. This sequence of events results in an implosion that heats up the target plasma to thermonuclear burn conditions in excess of two million degrees Celsius. The immense pressure created by the fusion burning the target plasma causes the plasma liner to first implode and then compress to a very high density, a process causing a thin layer of the liner to ignite and burn, thereby producing the fusion yield.

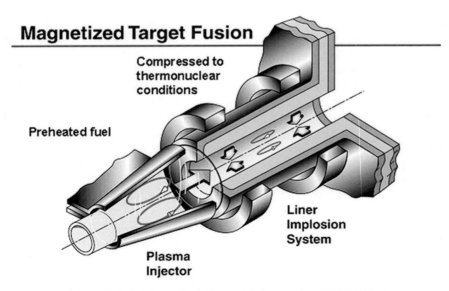

Figure 5.5 A Magnetized Target Fusion engine. (NASA.)

The outer layer of the liner carries hydrogen which moderates the neutrons created by the fusion reactions and, in the process, transfers most of the neutron energy to charged particles. The charged particles are then directed out, through a magnetic nozzle to produce thrust.

Magnetized target fusion today
Although MTF is a new entry into the space propulsion arena, it is a technology that promises low-cost development and rapid deployment using existing facilities and technology. The low system mass and volume, high Isp and thrust, combined with low thermal waste are features of MTF promising reduced trip times and increased payloads, obviously making the technology attractive to astronauts and mission planners. However, while MTF clearly surpasses more traditional approaches to propulsion and while many propulsion engineers argue fusion propulsion is inevitable, the technology will probably be not be realized in the near term. Nevertheless, MTF, in common with so many other high-tech concepts, represents a viable and realistic quest for advanced space propulsion. Developing these concepts, some of which previously belonged to the realm of science fiction, will always prove challenging but through research and aggressive technology development it will be possible to reduce the travel time to Mars to only a few weeks and, in the case of antimatter, to only a few days.

ANTIMATTER

Overview
One of the propulsion systems favored by many science fiction writers is antimatter, and for many rocket scientists, antimatter is viewed as the future of space travel. When it comes to fuels with high energy densities, antimatter/matter reactions can't be beaten since, when a particle and its antiparticle meet, they annihilate each other and their entire mass is converted into pure energy. Whereas several tonnes of chemical fuel are required to send humans to Mars, only a few tens of milligrams of antimatter would suffice. Unfortunately, this power comes at a price and more than just a few problems. For example, several antimatter reactions produce flashes of high energy gamma rays, capable of inflicting severe damage not only upon the human body, by penetrating tissue and making it radioactive, but also upon rocket engines, making them radioactive by fragmenting atoms of the engine material.

Antimatter concept of operations
Antimatter is sometimes referred to as the mirror image of normal matter because although it appears like ordinary matter, many properties are reversed. For example, normal electrons, the particles that carry electrical current, have a negative electrical charge whereas anti-electrons have a positive charge, which is why anti-electrons are often referred to as positrons. Positrons, the first known example of antimatter, were actually observed experimentally by scientist, Carl Anderson, in 1932, but it was

another twenty-three years before the positrons counterpart – the antiproton – was observed.

Problems with antimatter

One of the problems with antimatter is its rarity. A commonly held theory is that at the time of the Big Bang, antiparticles and particles were created in almost equal numbers, but if this were true, where is all the antimatter now? One answer is that particles outnumbered antiparticles, resulting in antiparticles being annihilated, but if this theory were true, there wouldn't be any antimatter left in the universe. Fortunately, this is not true, since low energy positrons are routinely used in a medical imaging technique known as Positron Emission Tomography (PET) in which positrons are produced as a result of the natural decay of radioactive isotopes. The problem with PET is that insufficient positrons are produced for the purposes of producing rocket fuel. However, high energy antimatter particles *are* produced in relatively large numbers at the world's particle accelerators.

Unfortunately, the current worldwide production of antimatter is only about one billionth of a gram per year and it is estimated that ten milligrams of positrons would be required for a manned mission to Mars. Using technology currently under development, producing just one milligram of antimatter per year would cost an estimated $10 billion, so from a practical standpoint, achieving antimatter propulsion is still pushing the limits of twenty-first century physics.

Antimatter spaceships

An antimatter spaceship would be fuelled by a positron reactor that would produce a specific impulse (I_{sp}) as high as five thousand seconds. Specific impulse is a measure of a rocket engine's efficiency, akin to kilometers per litre for a car engine. For example, the Space Shuttle Main Engine (SSME) has a maximum I_{sp} of about four hundred and fifty seconds, meaning one pound of fuel will produce a pound of thrust for four hundred and fifty seconds. By comparison, a nuclear reactor might generate an I_{sp} of nine hundred seconds. Thanks to the high I_{sp} of the positron reactor, the antimatter spaceship would transport astronauts to the Red Planet in less than seven weeks, compared to the four to six month journey using conventional chemical rocket technology. Unfortunately, building an antimatter spaceship will be no easy matter. First there is the problem of creating sufficient quantities of antimatter and then there is the problem of actually storing it in a small space. Storing antimatter represents a significant design challenge for antimatter spacecraft engineers since positrons annihilate normal matter so it isn't just a case of stuffing them in a container. Instead, antimatter particles must be contained by an electrical and magnetic field, requiring heavy magnets.

Fortunately, researchers think they may have found a way to store antimatter particles without surrounding rocket engines with layers of magnets. The solution is the Penning Trap (Figure 5.6), a super cold, evacuated electromagnetic bottle in which charged antimatter particles can be suspended. Since positrons are difficult to store using this method, antiprotons are stored instead. Scientists at Pennsylvania State University have already built a Penning Trap capable of holding ten million

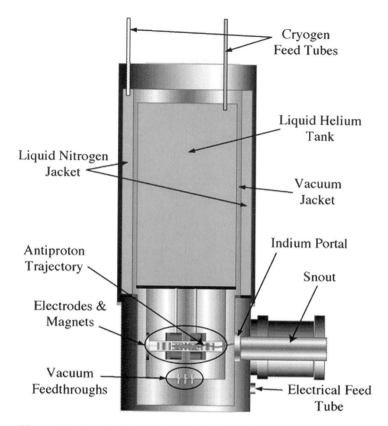

Figure 5.6 Penning Trap schematic. (Pennsylvania State University.)

antiprotons for a week, and are developing a second Trap with a capacity a hundred times greater.

One concept of an antimatter-fuelled manned Mars spaceship uses the Ion Compressed Antimatter Nuclear (ICAN) engine, developed by scientists as Pennsylvania State University. The ICAN engine would provide propulsive power through small nuclear explosions resulting from small one hundred and fifty gram pellets each the size of a golf-ball. Two hundred thousand of these pellets would be used as fuel in the ICAN engine and would explode at a rate of one per second to generate a continuous nuclear fusion reaction triggered by a microgram of antiprotons. The radiation produced by the fusion reaction would hit a hemispherical lithium hydride shield, protecting the crew. The heat from the exploding pellet would slowly ablate the shield at a rate of one micron per explosion and the resulting stream of atoms ablating from the surface of the shield would generate the thrust. The resulting blasts would propel the ship (Figure 5.7) through space, and massive shock absorbers would cushion the ship from the blasts. Using this propulsive power, the ICAN engine would produce an I_{sp} of 13,500 seconds and

Figure 5.7 An antimatter spaceship. (NASA.) See Plate 8 in color section.

be capable of accelerating the spaceship to one hundred kilometers per second. A byproduct of the propulsion would be radiation damage so the crew would be protected by a 1.2-metre-thick slab of lithium hydride shielding.

IN SUMMARY

This chapter has discussed advanced propulsion technology, ranging from the near-term visionary goal of Chang-Diaz's VASIMR^{TM} to the exotic research directed at realizing the potential of fusion and antimatter. The propulsion ploy in achieving short round-trip transits to and from Mars is to leave no stone unturned, whether investigating quantum vacuum energy or superluminal quantum tunneling. A near-term success in *any* propulsion technology may ultimately foster the space drive of tomorrow and lead to overcoming one of the major obstacles of deep space travel, namely, finding a fast and efficient way to get around. Through research and technology development, engineers are beginning to realize significant advances in space propulsion, which may ultimately lead to a time-saving system such as Chang-Diaz's VASIMR^{TM} being operational in time for the first manned mission.

REFERENCES

1. Borowski, S.K.; Dudzinski, L.A.; and McGuire, M.L. Bimodal Nuclear Thermal Rocket (NTR) Propulsion for Power-Rich, Artificial Gravity Human Exploration Missions to Mars. IAA-01-IAA.13.3.05. 2001.
1. Chang-Díaz, F.R.; Hsu, M.M.; Braden, E.; Johnson, I.; Yang, T.F. Rapid Mars Transits With Exhaust-Modulated Plasma Propulsion. NASA Technical Paper 3539, March 1995.
1. Chang-Díaz, F.R., and Yang, T.F. Design Characteristics of the Variable Isp Plasma Rocket. AIAA, 22nd International Electric Propulsion Conference, IEPC-91-128, October 1991.
1. Culver, D.W.; Kolganov, V., and Rochow, R. Low Thrust, Deep Throttling, US/CIS Integrated NTRE. 11th Symposium on Space Nuclear Power Systems. Albuquerque, New Mexico. January 9–13, 1994.

6

Mars hardware

The hardware described in this chapter is based on the architecture proposed by SpaceWorks Engineering Inc. (SEI) [6]. The SEI architecture (see Chapter 3) is similar to NASA's Constellation Design Reference Mission (DRM) and assumes a manned Mars mission will utilize the major elements of NASA's Constellation development effort including a new family of all-chemical launch vehicles, flight components, ground services and no in-situ resource utilization (ISRU).

The SEI mission plan was chosen to be the DRM for this book since it was judged to offer not only the greatest mass and performance advantage over other architectures, but also the most realizable and innovative architecture, thanks in part to the incorporation of the TransHab concept.

MARS MISSION ARCHITECTURE REVIEW

Before describing SEI's flight elements, it is useful to outline the overall concept of operations which commences with two Ares V launches. Departing December 2030, the two Ares V launch vehicles place a Trans-Mars Injection (TMI) stage, In-Space Propulsion Stage (ISPS), Earth Return Vehicle (ERV) and return TransHab into low Earth orbit (LEO).

In LEO, the TMI docks with the ISPS/ERV/Transhab configuration and departs for low Mars orbit (LMO). After TMI, the ISPS/ERV/TransHab stack undocks from the spent TMI stage, and solar arrays power the ISPS/ERV/TransHab during the Earth-Mars transit. Approaching Mars, the Mars Orbit Insertion (MOI) burn is performed by the ISPS, and the ISPS/ERV/TransHab remains in LMO until the return trip to Earth.

In April 2033, a third Ares V places a Mars Excursion Vehicle (MEV), comprising the Mars Surface Habitat (MSH) and its descent stage, directly on a TMI trajectory. Power for the MSH during the Earth-Mars transit is provided by solar arrays on a power and propulsion module attached to the MEV. When the MEV approaches Mars, an aerobraking/aerocapture maneuver is conducted, placing the MEV in LMO. From LMO, a terminal descent maneuver is performed, culminating in the descent stage and the MSH landing on the surface of Mars.

In the same month, the fourth Ares V launch places another MEV and the crew's Earth-Mars outbound TransHab in LEO. This MEV also contains an ascent stage, minimal crew habitat, descent stage, and a pressurized rover. This launch is followed by six crewmembers being launched into LEO by an Ares I. Once in LEO, the crew rendezvous and docks with the MEV/TransHab and the Orion is de-orbited and returns to Earth to be reused. On arrival at Mars, the crew transfers to the MEV ascent stage, jettisons the TransHab and lands on Mars, where they drive the pressurized rover to the site of the pre-deployed MSH.

After extended surface operations, the crew uses the MEV ascent stage to reach LMO, where they rendezvous and dock with the TransHab/ISPS/ERV configuration. During Mars-Earth transit, the crew lives in the TransHab and, on arriving in LEO, transfers into the ERV, which performs a direct Earth entry.

EXPLORATION SYSTEMS ARCHITECTURE STUDY

The Ares family of launch vehicles that will insert Mars mission assets into LEO will utilize technology derived from NASA's Space Shuttle. NASA's decision to choose Ares represented the outcome of careful consideration, study and evaluation of hundreds of commercial, government, and concept launch vehicle alternatives and architecture systems that could be used for human space exploration. When NASA performed these evaluations, it considered factors such as desired lift capacity, reliability, and the life-cycle development costs of different approaches. The culmination of these efforts was the release, in 2005, of the Exploration Systems Architecture Study (ESAS) [3], which analyzed architecture and launch vehicle requirements to provide the safest, most reliable, and cost effective system architecture to transport humans to Mars. Following exhaustive consideration and examination of Evolved Expendable Launch Vehicles (EELVs), Shuttle-derived vehicles and 'clean-sheet' architectures, it was determined, based on human safety, programmatic risk, reliability, and mission performance, that the architecture offering most advantages was one relying on Shuttle-derived technology.

The remainder of this chapter describes the design and development of NASA's new family of launch vehicles, the Crew Exploration Vehicle (CEV)/Orion destined to replace the Space Shuttle, and finally, the flight elements comprising the SEI Mars architecture.

ARES V

Design requirements
The design requirements of the hardware, software, facilities and personnel required to perform the Mars DRM are defined in NASA's 593-page Constellation Architecture Requirements Document (CARD). The CARD [2] is structured to provide mission planners and systems engineers with design guidance and an overview of the architecture's functional and performance requirements. An example

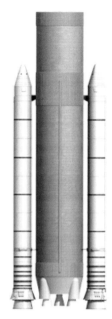

Core Stage
LOx/LH$_2$
5 RS-68 Engines
Al-Li Tanks/Structures

5-Segment
2 RSRB's

Figure 6.1 Ares V core stage. (NASA.)

of one of the CARD requirements for the Cargo Launch Vehicle (CaLV)/Ares V is presented in Table 6.1.

Ares V overview
Manufactured by the Michoud Assembly Facility (MAF) in Louisiana, the Ares V core stage (Figure 6.1), measuring 10m in diameter and 64 m in length, will be the largest rocket stage ever constructed and will be almost as long as the combined length of the Saturn V first and second stages, which were also manufactured by the MAF.

Table 6.1 CARD cargo launch vehicle description and requirements [2]

CaLV REQUIREMENTS

[CA0391-PO] The CaLV shall utilize twin shuttle-derived 5-segment SRBs along with a core stage that employs 5 modified RS-68 engines for first stage propulsion.
  Rationale: The CaLV will take advantage of the flight proven propulsion systems components developed for the Space Shuttle and Evolved Expendable Launch Vehicles (EELVs). These launch vehicle components, which have supported over 100 Space Shuttle and numerous Apollo missions, have extensive test/flight experience databases available for CaLV designers to leverage. In addition, CaLV designers will be able to leverage the ground processing/production facilities, workforce and tooling already in place to support Space Shuttle operations.

Ares V core stage propulsion

Powering the Ares V core stage will be a cluster of six Pratt & Whitney Rocketdyne RS-68 rocket engines (Figure 6.2), each capable of supplying 700,000 pounds of thrust. The RS-68 engine, the most powerful LOX/LH2 engine in existence, will be modified by a series of upgrades to meet NASA's standards. Providing the thrust for the mission elements is the J-2X engine (Figure 6.3), also powered by LOX/LH2. The J-2X is an evolved variation of the powerful J-2 upper-stage engine that powered the Apollo Saturn 1B and Saturn V rockets to the Moon.

During the Ares V missions, the first time the J-2X will be ignited will be 325 seconds after launch at an altitude of approximately 120km, following separation of the Ares V first stage from the flight elements. Following a burn of 442 seconds, during which the J-2X will use 131,818 kilograms of propellant, the flight elements will be placed in LEO. The second ignition of the Ares V J-2X will occur once Orion, delivered to LEO by the Ares I, has docked with the MEV and outbound TransHab. Once these flight elements are mated the J-2X will provide sufficient power to accelerate the mated vehicles to the escape velocity required for the Orion/MEV/TransHab configuration to break free of Earth's gravity and enter a trajectory known as trans-Mars injection (TMI).

ARES I

Design history

The evolution of Ares I can be traced back to 1995, when Lockheed Martin submitted a report of work conducted under an Advanced Transportation System Studies (ATSS) contract [1] to the Marshall Space Flight Center (MSFC). The summary of the ATSS report described launch vehicle configurations similar to the current Ares I design, featuring liquid rocket second stages stacked above Shuttle Solid Rocket Booster (SRB) first stages and variants of these designs that included the possible use of the J-2S engine for the upper stage.

In January 2004, nearly nine years after the ATSS report, President Bush announced the Vision for Space Exploration (VSE), resulting in NASA chartering the ESAS to determine the requirements and configurations for crew and cargo systems for exploration missions to the Moon and Mars. This task resulted in the selection of a Shuttle-derived launch architecture which initially envisaged using a standard four-segment SRB for the first stage and a SSME variant for the upper stage. However, although the initial design was approved, it was quickly determined the Orion CEV would be too heavy for the planned 4-segment launch vehicle, a finding that spurred NASA to reduce the size of Orion and add an extra segment to the SRB first stage. Also, rather than use the single SSME as suggested in the original design, the revised option chose to use the Apollo-derived J-2X engine. This choice was due in part to the latter engine's advantage of being able to start in mid-flight and in a vacuum, whereas the SSME was designed to start only on the ground.

M-RS68-07/06/00-ST001

Figure 6.2 RS-68 rocket engine. (NASA.)

Figure 6.3 J-2X rocket engine. (NASA.) See Plate 9 in color section.

Design endorsement

In October 2006, NASA announced it would extend a previous contract with ATK Thiokol of Brigham City, Utah, to continue design and development of the Ares I first stage. Under the contract, ATK Thiokol was required to maintain a Design Development Testing and Evaluation (DDT&E) schedule for the first stage, to expedite procurement of new nozzle hardware, and maintain design and engineering analysis, leading to a Systems Requirements Review (SRR). The SRR (see Glossary), completed on 4th January, 2007, confirmed Ares I system requirements had been validated and fulfilled mission requirements, and it duly endorsed the Ares I architecture design concept. The following day, NASA extended the Ares I development contract with ATK Thiokol. Under this contract action, ATK Thiokol was tasked with maintaining design and engineering analysis, to prepare pilot parachute development tests, and to support the initial test flight scheduled for the spring of 2009.

Ares I design

In common with the design requirements of Ares V and Orion, the constraints and specifications concerning the development of Ares I are described in the CARD (Table 6.2).

Table 6.2 CARD crew launch vehicle/Ares I requirements [2]

CLV DESCRIPTION

The CLV is the launch vehicle for Orion. It consists of a 5-segment solid rocket booster first stage and a cryogenic liquid hydrogen/oxygen fuelled upper stage consisting of a structural tank assembly and a J-2X engine. The first stage is reusable and the upper stage is discarded after Orion has separated during ascent.

CLV REQUIREMENTS

[CA1065-PO] The CLV shall limit its contribution to the risk of loss of mission (LOM) for any mission to no greater than 1 in 500.
  Rationale: The 1 in 500 means a 0.002 (or 0.2%) probability of loss of CLV mission for any Constellation DRM. This requirement is driven by CxP 70003-ANX01, Constellation Need, Goals, Objectives, Safety Goal CxP-G02: provide a substantial increase in safety, crew survival and reliability of the overall system over legacy systems.

[CA0389-PO] The CLV shall use a single 5-segment Solid Rocket Booster, modified from the Space Shuttle Solid Rocket Booster (SSRB), for first stage propulsion, and a single modified Apollo J-2X engine for second stage propulsion.
  Rationale: The CLV will take advantage of the flight-proven propulsion system components developed for the Space Shuttle and Apollo. These launch vehicle components, which have supported over 100 Space Shuttle and numerous Apollo missions, have extensive test/flight experience databases available for CLV designers to leverage. In addition, CLV designers will be able to leverage the ground processing/ production facilities, workforce and tooling already in place to support Space Shuttle operations.

Ares I first stage design
The Ares I first stage is a five-segment RSRB which will burn a specially formulated and shaped solid propellant called polybutadiene acrylonitrile (PBAN). Above the first stage (see Figure 3.11 in color section) sits a forward adapter/interstage that interfaces with the Ares I liquid-fuelled upper stage. Above the interstage is a forward skirt extension, housing the Main Parachute Support System (MPSS) and main parachutes for recovery of the first stage. The frustrum, located at the top of the first stage elements, provides the physical transition from the smaller diameter of the first stage and the larger diameter of the upper stage. Other modifications made to the original Shuttle SRB included removing the ET attachment points and altering the propellant grain inside the SRB.

Ares I upper stage design
The Ares I upper stage being designed by MSFC is based on the internal structure of the Space Shuttle's ET. The upper stage was originally designed to incorporate separate fuel and oxidizer tanks separated by an intertank (a structure used in the

Apollo program) but, due to mass restrictions, engineers decided instead to use a common bulkhead between the tanks.

The upper section of the upper stage includes a spacecraft adapter (SCA) system designed to mate with Orion, while the lower section includes a thruster system to provide roll control for the first and upper stage. Power for the upper stage is provided by the LH2 and LOX-fuelled J-2X rocket engine.

Ares I avionics

The Instrument Unit Avionics (IUA) provides guidance, navigation and control for the Ares I. The IUA includes subsystems such as the J-2X engine interface, upper stage reaction control system (RCS), first stage roll control system, Hydraulic Power Unit Controller (HPUC), Data Acquisition/Recorder Unit and the Ignition/Separation Unit. Power for all the subsystems is provided by the Electrical Power System (EPS), comprising batteries, power distribution and control units, DC-to-AC Inverter Units and cabling. Located in the IUA, the EPS ensures redundant sources of 28-volt direct current (VDC) from the time ground power is removed, prior to launch, until the end of the mission. In addition to the primary EPS, the upper stage also features an independent EPS, located in the Interstage, whose purpose is to provide power to the First Stage Roll Control System (FSRCS) thrusters.

Ares I safety systems

To prevent combustible LOX and LH2 accumulating to dangerous levels while Ares I is on the launch pad, the upper stage is fitted with a Purge System. The Purge System ensures thermally conditioned inert nitrogen gas is pumped into the closed compartments of the upper stage, while at the same time exhausting excess nitrogen through special vents at the bottom of the compartments. Another safety system used on the launch pad is the Hazardous Gas Detection System (HGDS), which samples, detects, and measures the concentration of hazardous gases in the compartments of the upper stage prior to launch.

Ares I test flights

When operational, Ares I will utilize a five-segment booster but during the first suborbital test flight, scheduled for June 2009, a four-segment booster will be used as the five-segment variant will not be ready. Instead, NASA will mount a fifth dummy segment on top of the standard four-segment version to replicate correct aerodynamic, mass and CG properties of the final design. Positioned above the booster will be the pilot, drogue and main parachutes, protected by an Aero Shell cover, shielding the parachutes from the damaging blast of the upper stage engine ignition.

Nominal mission profile

When operational, a typical mission will commence with the Ares I first stage booster lifting Orion to an altitude of 50 km, at which point the first stage will separate and fall back to the ocean to be recovered in the same manner as Shuttle SRBs are recovered today. The Ares I upper stage will then take over and carry Orion to an elliptical orbit of 245 km. Once the upper stage has separated, Orion's propulsion system will power the spacecraft to its 300-km circular orbit. There, Orion will dock with the MEV/TransHab stack and depart for LMO.

Test flights

NASA has learned through experience that operating costs can be reduced by thorough and incremental testing of its launch vehicles. This "test as you fly" philosophy was implemented during the Saturn development program, which conducted multiple demonstration and verification flight tests before finally certifying the vehicle safe for humans.

A different tactic was adopted for the Space Shuttle program which involved much less testing but did not appreciably reduce schedule costs. This is why NASA has reverted to the Saturn approach and plans to conduct a progressive series of demonstration, verification, and mission flight tests before certifying Ares I safe for humans. The first of these flights will take place in 2009, with the launch of the Ares I Flight Test Vehicle (FTV) on a suborbital development flight test.

Development problems

"I hope no one was so ill-informed as to believe that we would be able to develop a system to replace the shuttle without facing any challenges in doing so. NASA has an excellent track record of resolving technical challenges. We're confident we'll solve this one as well."

NASA Administrator, Dr. Michael Griffen,
Associated Press, 20 January, 2008

Thrust oscillation

Perhaps the most troublesome performance matter concerning engineers is the issue of *thrust oscillation* (TO), a phenomenon also known as *resonant burning*. TO is a solid rocket motor (SRM) occurrence, characterized by increased acceleration pulses that may occur in the latter part of the first stage phase of the flight. The pulses, which may vary in amplitude to the point at which structural damage is a possibility, are generated as a result of vortices created by burning propellant inside the SRM. The vortices may, as a result of flow disturbances, coincide with acoustic modes of the combustion chamber, thereby generating longitudinal forces which may in turn increase loads encountered by Ares I during ascent to orbit. According to NASA's computer-modeled performance of the Ares I first stage, noticeable TO would occur at lift-off and would become progressively more severe, eventually peaking at 110 seconds into flight. If a certain frequency and amplitude of TO were to set up a specific resonance, worst-case vibration loads of between 4.5 and 5.5 g would be experienced by the crew. Such a frequency would harm not only critical components such as the guidance, navigation, and control avionics but also completely incapacitate the crew.

Similar TO effects were noticed during the Shuttle SRB testing and during some Shuttle flights but there is insufficient information (development flight instrumentation was only fitted for Shuttle flights STS-1 through STS-4) for the Ares I 'tiger team' to foresee exactly the how the oscillation predicted for the Ares I FS might affect the LV and its crew.

Although the TO problem initially appeared to be a potentially destructive event, after analyzing the problem, the Ares I 'tiger team' suggested TO may not be as severe as initially reported by space media analysts. To remedy the TO problem, engineers will install an active tune mass absorber, designed to detect the frequency and amplitude of the TO with accelerometers, and use battery-operated motors to move weights up and down to damp it out. This fix should reduce any vibration loads to less than 0.25g, which is considered low enough for astronauts to read displays and react to changing conditions effectively.

ARES I AND V PROPULSION

Propulsion systems often represent the most technologically challenging and risky aspect of any launch vehicle development due to the extremely high operating temperatures, pressures, and mixture ratios involved. To reduce these risks NASA opted for the J-2X engine, a choice resulting from recommendations made by the original ESAS study.

J-2 and J-2X history

The J-2 engine was originally developed by Rocketdyne in the 1960s for the Saturn IB and Saturn V. A simplified version of the engine was also designed by Rocketdyne and designated the J-2S, which, due to design problems, was never flown and not recommended for further development. However, variants of the J-2 design were considered as options for further study and it was an advanced derivative of one of these J-2 variants that resulted in the final J-2X design.

At approximately $25 million per engine, the J-2X will cost less than its $55 million SSME counterpart. However, despite the lower cost, the J-2X has a performance advantage over the SSMS since, unlike the SSME, which was designed to start on the ground, the J-2X will feature an air-start and vacuum-start capability, enabling Ares I to fly the direct-insertion profile demanded by the mission architecture.

J-2X concept of operations

The J-2X fitted to Ares I will ignite approximately 126 seconds after lift-off, following separation of the vehicle's first stage at an altitude of 60km. The engine will burn for 465 seconds, expending 137,363 kilograms of propellant and lift Ares I to an altitude of 128 km. Following engine cut-off, Orion will separate from the upper stage, whereupon Orion's engine will ignite to provide the propulsive power to insert the capsule into LEO. The Ares I upper stage with the J-2X engine attached will then de-orbit to splash down in the Indian Ocean. Neither flight element is designed to be reused.

J-2X hardware

The turbomachinery of the J-2X, comprising primary power-generating elements such as the gas generator, turbopumps, valves, and feed lines, will utilize hardware initially developed for its sister engine, the J-2S. The operation of this type of rocket engine requires the turbopumps to supply LOX and LH2 to the main combustion chamber, where the fuel and oxidizer mix and burn at extremely high pressures and temperatures to produce gas, which is forced through the nozzle to produce thrust. The components of the J-2S engine have already been subject to extensive and rigorous testing as part of NASA's 1990s aerospike engine development program which was tasked with developing the engine for the X-33 single stage-to-orbit (SSTO) reusable launch vehicle (RLV).

ORION

Concept of operations
SEI's Mars mission utilizes what NASA refers to as a Block 3 Orion variant to transfer three crewmembers between Earth and LEO and from LEO to Earth. In SEI's mission profile the Block 3 Orion is launched by Ares I into an orbit matching the inclination of the awaiting MEV/TransHab, launched previously by the Ares V. Orion is first injected into a 55 by 296 km altitude orbit while the MEV/TransHab loiters in a circular orbit of 800 to 1,200-km altitude. It then takes Orion up to 2 days to perform orbit-raising maneuvers to close on the MEV/TransHab, before conducting a standard ISS-type rendezvous and docking approach to the stack. After docking, the crew performs a leak check, equalizes pressure with the MEV/TransHab, and opens hatches. Once crew and cargo transfer activities are complete, Orion returns to Earth for reuse. This particular aspect of the architecture constitutes one of the mission performance advantages over the NASA plan in which Orion is configured to a quiescent state and remains docked to the flight elements for the trip to Mars and back.

On the return Orion is attached to the TransHab as it approaches Earth. Shortly before re-entry, the crew performs a pre-undock health check of all entry critical systems, transfers to the Orion, closes hatches, performs leak checks, and undocks from the TransHab, which is then discarded. The Orion then conducts an onboard-targeted, ground-validated burn to target for the proper entry corridor and, as entry approaches, the Orion maneuvers to the proper Entry Interface (EI) attitude for a direct-guided entry to the landing site. The Orion then performs a nominal landing at the primary land-based landing site and the crew and vehicle are recovered.

Orion module overview
Orion (Figure 6.4), which is being built by Lockheed Martin, comprises four functional modules. On top of the Ares I launch vehicle sits the Spacecraft Adapter (SA) which serves as the structural transition to the Ares I launch vehicle. Above the SA is the unpressurized Service Module (SM) which provides propulsion and electrical power to the pressurized Crew Module (CM)/Orion. Above Orion sits the Launch Abort System (LAS), providing the crew with an emergency escape system during launch. Although the modular design may appear familiar to those who remember the Apollo program, the vehicle configuration takes full advantage of all the technology the 21st century has to offer.

Figure 6.4 Exploded view of Orion showing launch escape system, capsule shroud, service module and interstage. (NASA.)

Designing Orion

Orion design
Orion's design is guided by similar CARD objectives, requirements and constraints to those governing the design of Ares I and V (Table 6.3).

Table 6.3 CARD Orion/CEV description and requirements [2]

ORION/CEV DESCRIPTION

The Orion/CEV System consists of a Crew Module (CM), a Service Module (SM), a Launch Abort System (LAS), and a Spacecraft Adapter (SA), and transports crew and cargo to orbit and back. The CEV system will be used in all phases of the Constellation program. Initially, Orion transports crew and cargo to and from the ISS and an uncrewed configuration transports pressurized cargo to and from ISS. It will subsequently transport crew and cargo to and from a lunar orbit for short and extended duration missions. Finally, Orion or a derivative will support a mission to a Mars transfer vehicle, and then return the crew and cargo to Earth after separation from this vehicle.

ORION/CEV REQUIREMENTS

[CA0324-PO] Orion shall return to Earth on land at designated CONUS[1] landing sites.
Rationale: Returning to land at designated CONUS landing sites reduces risk and cost by reducing necessary recovery force assets, increasing proximity to U.S. medical facilities, increasing security, ensuring a prepared landing site free of hazards and supporting vehicle reuse.

Table 6.3 *cont.*

[CA0448-PO] Orion, when operated by the crew, shall be controllable by a single crewmember.

    Rationale: vehicle systems must be designed so that more than one crewmember is not required to operate the vehicle. There may be circumstances where crewmembers are unconscious or incapacitated, leaving only a single crewmember capable of vehicle control.

[CA3142-PO] Orion shall perform navigation and attitude determination during all mission phases including pre-launch.

    Rationale: Navigation and attitude determination are required onboard Orion to accomplish mission critical activities such as communications antenna pointing, maneuver execution and performance monitoring, entry guidance and docking. Navigation may include maintenance of a ground uploaded vehicle state or updates of the vehicle state by processing data from onboard sensors. All mission phases include pre-launch activities through touchdown and recovery, including aborts, even when Orion is not controlling the vehicle.

As with the design of Ares V and I, the CARD requirements provide mission planners and engineers with reference points to be used when defining the optimum vehicle design to satisfy the overall mission goal. This process generates design concepts satisfying mission requirements, often achieved through competition between contractors. Once options have been selected for further development, trade studies are conducted to support the decisions, before developing conceptual designs to evaluate feasibility and cost. At this stage, models of the vehicle may be built in order to define vehicle subsystems, mass, and size considerations. Finally, after comparing conceptual designs, the design may be iterated or a baseline will be selected for further development.

Next, the design team considers what are known as *design drivers*, usually grouped under the four phases of the mission, namely *Earth ascent*, *space transfer*, *descent and ascent*, and *Earth re-entry*. For example, a design driver for the Earth ascent phase requires engineers to consider crew orientation during launch and the type of emergency egress available to the crew, whereas for the Earth-Mars transfer phase the design team must consider factors such as staging options, radiation protection, and provisions for rendezvous and docking. Using the design drivers and CARD's Orion requirements as reference guidelines, the team then proceeds with designing the vehicle.

Orion systems and subsystems

Vehicle overview
Before describing the systems and subsystems of the Mars Orion variant, it is useful to have an overview of the mission requirements demanded of the vehicle. In addition to providing habitable volume and life support for the crew during the ascent and entry phases of the mission, Orion also features a docking capability and a means of transferring the crew to the MEV/TransHab. On return to Earth, a thermal protection

system (TPS) protects the crew during atmospheric entry and a combination of parachutes and four Kevlar airbags are deployed to protect the crew during a land touchdown, although a water flotation system may also be deployed in case of a water landing. Upon recovery Orion is refurbished and made ready for its next mission.

Vehicle shape

Orion's shape was largely a consequence of the familiar triad of cost, speed of design and safety, resulting in the decision to use a blunt-body capsule. One of the main advantages of the blunt-body design lay in familiarity with the aerodynamic design which, thanks to the experience of Apollo, resulted in ascent, entry, and abort level loads familiar to engineers and, consequently, in reduced design time and cost. However, the shape is about the only design aspect Orion shares with its illustrious predecessor as it provides a much larger habitable volume and incorporates the very latest in avionics and life support technology (Table 6.4).

Table 6.4 Orion configuration and mass summary

Configuration Summary				Mass Properties Summary	
Diameter	**5 m**	Total Delta V	**164 ft/s**	Dry Mass	**7907 kg**
Pressurized Volume	**691.8 ft**	RCS Engine Thrust	**100 lbf**	Propellant Mass	**175 kg**
Habitable Volume	**361 ft³**	Lunar Return Payload	**220 lbs**	Oxygen/Nitrogen Mass/Water	**128 kg**
				GLOW	**8502 kg**

Vehicle materials

Orion's pressure vessel structure uses Aluminum 2024 honeycomb sandwich for the face sheets and Aluminum 5052 for the honeycomb core, a combination of materials enabling the vehicle to withstand 14.7 psia internal cabin-pressure. For the purposes of rendezvous and docking operations, Orion is fitted with five double-paned fused silica windows, two forward facing, two side windows, and a fifth window located within the side ingress/egress hatch.

Orion's Outer Mold Line (OML) is composed of graphite epoxy/Bismaleimide (BMI) composite skin panels, whose structure provides the vehicle's aerodynamic shape and serves as the attachment assembly for the Thermal Protection System (TPS).

Vehicle thermal protection

Orion's TPS must have the capability of shielding the crew from the thermal environment encountered during ascent, ascent abort, on-orbit operations, and re-entry. One of the most important factors engineers consider when deciding which material to use is the TPS heating rate for different entry trajectories, a design aspect reflected in the placing and thickness of the material.

Figure 6.5 Orion prototype/manufacturing demonstration unit. (NASA.) See Plate 10 in color section.

The thermal material used for Orion's TPS is Phenolic Impregnated Carbon Ablator (PICA), an ablative material that will protect Orion from the high heat flux generated during re-entry. As Orion enters the atmosphere, the extreme heat will cause the PICA heat shield material to chemically decompose, a process known as *pyrolysis*. As PICA is pyrolyzed, the material will begin to char, before melting and finally sublimating, creating a cool boundary layer to protect the vehicle.

In January 2008, Orion's prototype heat shield underwent testing at NASA's Kennedy Space Center (KSC). The prototype (Figure 6.5), referred to as a manufacturing demonstration unit (MDU) has the same dimensions as the actual heat shield that will protect Orion as it re-enters Earth's atmosphere on its return from Mars. The MDU currently rests in Hangar N at Cape Canaveral Air Force Station where it is undergoing several months of nondestructive laser scan and X-ray evaluation testing, the results of which will be used to construct the actual heat shield. The actual heat shield will consist of almost two hundred PICA blocks that share the same delicate characteristics as the tiles on the Space Shuttle. Unlike the Space Shuttle however, Orion's ablative TPS is designed to be used only once.

Vehicle propulsion

Orion's propulsion system comprises a Reaction Control System (RCS) which includes a number of elements such as the RCS tanks, the RCS pressurization system, the primary RCS thrusters, and the back-up RCS thrusters and RCS tanks.

The RCS enables the Orion to perform exoatmospheric maneuvers and to orient itself during atmospheric re-entry. It also provides astronauts with a means of counteracting induced spin and dampening induced pitch and yaw instabilities. In the event of a loss of power, the Orion has a fully independent back-up RCS.

The propellant for the RCS is a bipropellant system comprising Gaseous Oxygen (GOX) and liquid ethanol. The GOX mixture, which also feeds the life support system (LSS), is stored in four cylindrical graphite-composite Inconel tanks mounted at the base of the vehicle, whereas the liquid ethanol is stored in two similar tanks and is pressurized by means of a high-pressure Gaseous Helium (GHe) system. Together, the GOX and ethanol mixtures feed the twelve 445 N thrusters arranged in pairs to thrust in the pitch, roll, and yaw directions.

Vehicle power

Three primary rechargeable Lithium-ion batteries, 28-Volt Direct Current (VDC) electrical power buses, power control units (PCUs), and back-up batteries comprise Orion's power subsystem, providing primary electrical power and distribution, and energy storage.

The three primary Lithium-ion batteries, capable of providing 13.5 kW-hr, were chosen for the high amount of energy they provide, good charge retention, volume, and low drain life. If more energy is required there is a fourth battery, providing one level of redundancy capable of providing 500 W of 28-VDC power for forty-five minutes. The Lithium-ion batteries and the two distinctive 'Mickey Mouse' solar arrays, feed electrical power to Orion's power distribution system. The power generated by these two systems results in an arrangement capable of distributing 28 VDC to the vehicle via a power distribution system comprising a complex arrangement of jumper cables, brackets, cable ties, and primary and secondary distribution cables and avionics wiring.

Vehicle communications

Command and Control (C&C) over all of Orion's operations is provided by the avionics subsystem comprising Command, Control, and Data Handling (CCDH), guidance and navigation, communications, and cabling and instrumentation. The CCDH's role is to process and display important spacecraft data to the astronauts on multifunction liquid crystal displays (LCDs) and control panels. A part of the CCDH system includes two sets of translational/rotational/throttle hand controllers, enabling the crew to take manual control of the vehicle when required.

The equipment providing the crew with on-orbit and in-transit vehicle attitude information, vehicle guidance and navigation processing information is supplied by the guidance and navigation system. The heart of the system is the Global Positioning System (GPS)/Inertial Navigation System (INS) that works in conjunction with two star trackers, video guidance sensors and two Three-Dimensional scanning Laser Detection and Ranging (LADAR) units.

Orion's communications components include S-band/Search and Rescue Satellite-aided Tracking (SARSAT), Ultrahigh Frequency (UHF) communications, information storage units and a high rate Ka-band communications system.

Orion's avionics

Orion features a start-of-the-art flight control system consisting of three briefcase-sized Honeywell Flight Control Modules. Two of the control computers could completely fail and the third will still be able to fly the vehicle. In the worst-case scenario of a complete power failure, Orion carries an emergency system powered by independent batteries providing the crew with enough capability to bring the vehicle back safely. In common with fighter aircraft cockpits, Orion's control systems rely heavily on *sensor fusion*, a type of automation, relieving the astronaut-pilot of being a sensor integrator and allowing him/her to focus on the mission. Such a system makes sense, given that many astronaut-pilots came from advanced cockpits such as the F-15 and F-22.

In Orion's cockpit, astronaut-pilots will be able to change displays as if on a revolving panel, thanks to four flat-screen displays, each about the size of a large desktop monitor. During ascent to orbit, the displays will operate similarly to the screens in conventional airliners. One display will show an artificial horizon, another will display velocity and a third will show altitude. The fourth display will show life support status and communications information. Once Orion reaches orbit, the displays will change to readouts showing rendezvous and docking information such as the vehicle's flight path, range, and rate of closing to the Mars flight elements.

Environmental control and life support system

Orion's Environmental Control and Life Support System (ECLSS) includes all the items necessary to sustain life and provide a habitable environment for the crew. In addition to the equipment for storage of nitrogen and oxygen, the ECLSS also ensures the atmosphere inside Orion remains contaminant-free and provides fire detection and suppression capability.

The nitrogen gas required to sustain three crewmembers is stored in cylindrical graphite composite Inconel 718-line tanks, while the oxygen is stored in the four primary RCS oxygen tanks. Atmosphere regulation is provided by a combined Carbon Dioxide and Moisture Removal System (CMRS), which ensures carbon dioxide levels are regulated, and an ambient temperature catalytic-oxidation (ATCO) system for contaminant control. Fire-detection and suppression capability consists of standard spacecraft smoke detectors and a fixed Halon fire-suppression system, identical to the ones used on the ISS. Other ECLSS components include cabin fans, air ducting, and humidity condensate separators, most of which are based on existing Shuttle technology or ISS systems.

Potable water is stored in four spherical metal bellows tanks, each capable of holding 53 kg of water. Once again, these tanks are similar to the ones installed on the Shuttle. In the event of a contingency EVA, requiring all crewmembers to don EVA suits and the cabin to be pressurized, the ECLSS includes the necessary umbilicals and ancillary equipment.

Active thermal control system

Orion's hardware will give off a lot of heat during operation. Since the behavior of fluids and the process of heat transfer are very different in microgravity, spacecraft engineers designed special equipment to radiate Orion's excess heat into space. To ensure a comfortable environment for the crew, NASA's Glenn Research Center (GRC), in partnership with the Jet Propulsion Laboratory (JPL) and Goddard Space Flight Center (GSFC), has developed the Active Thermal Control System (ATCS). The ATCS provides a temperature control capability for the vehicle, consisting of a propylene glycol/water fluid loop with a radiator and fluid evaporator system. The fluid loop works as a heat rejection system by using cold plates for collecting waste heat from the equipment, while a cabin heat exchanger regulates atmosphere temperature.

To deal with high heat loads, the ATCS includes a dual-fluid evaporator system that works by boiling expendable water or Freon R-134A in an evaporator. This cools the heat rejection loop fluid which is circulated through the walls of the evaporator, which in turn causes vapor to be generated and then vented. The reason for a dual-fluid system is because water does not boil at the ATCS fluid loop temperatures and pressures at 30,000 m altitude or less, meaning that from the ground to an altitude of 30,000 m Freon R-134A is used.

Crew living area

Although Orion provides a larger habitable volume than the Apollo vehicle, crew accommodations are fairly snug. Many of the facilities such as the galley are identical to those onboard the Shuttle, astronauts waiting to rendezvous and dock with the MEV/TransHab will still be eating freeze-dried and irradiated prepackaged foods. There is little in the way of improvements in the toilet department, a passive Mir-style design, comprising a privacy curtain, contingency waste collection bags and what NASA euphemistically refers to as a suitable user interface!

Non-propellant

Oxygen, nitrogen, potable water, FES water and Freon constitute Orion's non-propellant components. Oxygen is used for breathing by the crew but also for contingency EVA consumption and in the event of rapid or explosive decompression, requiring full-cabin repressurization. Nitrogen is used for the cabin atmosphere and also for waste management and the carbon dioxide regeneration system.

Parachute and landing system

Orion's return to Earth will be achieved through a water landing near San Clemente Island, northwest of San Diego. However, in the event of an off-nominal/abort contingency event Orion will still be capable of a land landing thanks to a wraparound partial airbag system comprising four cylindrical airbags located on the "toe" of the capsule.

Orion's parachute system is packed between the vehicle's pressure vessel and OML, which is near the docking mechanism. Consisting of two 11m diameter drogue parachutes (Figure 6.6) and three round primary parachutes (Figure 6.7), each with a diameter of 34m, the parachute system ensures a nominal landing speed of 8 m/s with all three parachutes deployed and a landing speed of 8.5m/s with one failed

Figure 6.6 Orion's drogue chutes. (NASA.)

Figure 6.7 Orion's primary chutes. (NASA.)

parachute. The system is automated much like the automatic opening device (AOD) used to open civilian parachutes, which deploy canopies using a sensor that detects a dynamic pressure. In the case of Orion's drogue system, the dynamic pressure is dialed in to 7,000 m altitude and a 400 km/h sink rate. The drogue chutes slow the sink rate to 200 km/h at an altitude of 3,300 m, the deployment altitude for the primary system.

Figure 6.8 Ares I Launch Abort System. (NASA.)

In the event of a land landing, Orion is cushioned by four inflatable Kevlar airbags. As the vehicle descends the airbags are deployed out of the lower conical backshell. Two panels jettison, permitting the airbags to inflate and wrap around the low hanging corner of the heat shield to provide energy attenuation upon landing. Once Orion has landed, the airbags vent at a specific pressure to facilitate a controlled collapse rate.

Launch abort system

The Launch Abort System (LAS) is designed to pull Orion away from the thrusting Ares I first stage in the event of an off-nominal situation (Figure 6.8). Compared to many other mission elements, the LAS is a relatively simple design, sharing many features with the Apollo Launch Escape System (LES). The system incorporates an active tractor design utilizing a canard section below the attitude-control motor element. Below the canard section are four jettison motors sitting atop the systems interstage. Below the interstage is the abort motor element, comprising four exposed, reverse-flow nozzles, and attached to the aft end of the abort motor element is the adapter cone, which in turn is attached to the boost protective cover (BPC) that essentially covers Orion. Following second stage ignition, the LAS is discarded, after which any contingency for abort is provided by the SM propulsion system.

Orion abort modes

[CA0466-PO] Orion shall perform aborts from the time the Orion abort system is armed on the launch pad until the mission destination is reached.

Rationale: Abort at any time is part of NPR 8705.2, Human rating Requirements for Space Systems, as well as the program policy on crew safety. This Orion requirement will cover all of the flight phases from abort system arming on the launch pad through docking with the ISS or LSAM landing. Orion must be capable of supporting an LSAM Descent abort and subsequent redocking. After reaching the destination, all other scenarios are covered by the return capabilities.

The CARD states there shall be no period during a mission in which a survivable abort mode is not available to the crew. The CARD also requires that no abort mode shall land Orion in the North Atlantic Ocean more than 250 km from St. John's, Newfoundland or Shannon, Ireland, an area also known as the North Atlantic Downrange Abort Exclusion Zone (DAEZ). These stipulations have required engineers to perform crucial analysis of propellant loading, since Orion's modes of abort are driven by the relatively low lift capability of Ares I. Engineers must also consider the delta-V capabilities that may allow a crew to abort-to-orbit (ATO), instead of a landing in the North Atlantic or possibly in the continental United States (CONUS).

To determine abort capabilities, engineers assessed two propellant loadings, one which maximized the earliest Targeted Abort Landing (Max-TAL), and one which maximized the earliest ATO capability (Max-ATO). For each of the propellant loadings, engineers calculated five abort-thrust magnitudes (7500 lbf, 8300 lbf, 10,000 lbf, 10,800 lbf, and 10,980 lbf), each magnitude representing various combinations of main engine and auxiliary engine thrust levels. For the purposes of calculation, the specific impulse of 323 seconds remained the same. Next, engineers identified three abort modes concerned with system-design impacts of the North Atlantic DAEZ requirement, the first of which was designated the Untargeted Abort Splashdown (UAS).

In the event of an UAS (Table 6.5), resulting in a water recovery off the Atlantic coast of either the United States or Canada, the SM's RCS fire to separate Orion from the Ares US, after which astronauts fire Orion's thrusters to control and guide the vehicle's bank angle during re-entry, being very careful to avoid acceleration levels that might subject them to excessive G-loads.

The next abort mode engineers analyzed was a Targeted Abort Landing (TAL), requiring the crew to first use the SM RCS to separate from the Ares I US and then fire the OME to boost the landing area to within 250 km of Shannon, Ireland. This type of abort (Table 6.5) is available once the SM propulsion system is capable of ensuring the required delta-V, while maintaining a minimum altitude of at least 121,200 meters to avoid excessive heat on re-entry.

The ATO abort mode uses the SM RCS to separate Orion from the Ares I US, and uses the OME to increase the apogee altitude to approximately 160 km, after which Orion coasts to apogee and performs an insertion burn to ensure a stable

Table 6.5 Orion abort modes[1]

Abort Mode	Phase	Description
Untargeted Abort Splashdown (UAS)	Abort Initiation	Abort is initiated at t_0
		Orion's state is assigned based on interpolated CLV at t_0
		Orion coasts to re-entry interface.
	Re-entry	CM separates from SM at re-entry interface.
		Initial pitch angle interpolated from trimmed aero-dynamic database based on Mach number.
		CM re-enters atmosphere. Bank angle optimized so abort initiation may occur as early as possible while still landing within 250 km of St. John's International Airport.
Targeted Abort Landing (TAL)	Abort Initiation	Abort is initiated at t_0
		Orion's state is assigned based on interpolated CLV at t_0
	Separation	CM separates from SM and drifts for 15 sec.
	Main Engine Burn	Main engine and auxiliary thrusters fired to boost downrange landing point into TAL recovery area, while maintaining altitude limitation of 121,200 m.
	Re-entry Interface	Orion coasts to re-entry interface 90,900 m.
	Re-entry	Orion re-enters atmosphere, deploys parachutes at 15,150 m. and lands near TAL recovery area.
Abort to Orbit (ATO)	Abort Initiation	Abort is initiated at t_0
		Orion's state is assigned based on interpolated CLV at t_0
	Separation	CM separates from SM and drifts for 15 sec.
	Main Engine Burn	Main engine and auxiliary thrusters fired to boost Orion's apogee altitude to 160 km.
	Coast to Apogee	Orion coasts to almost apogee.
	Circularization	Orion circularizes using main engine.

1. Adapted from: Falck, R.D., Gefert, L.P. Crew Exploration Vehicle Ascent Abort Trajectory Analysis and Optimization. Glenn Research Center, NASA/TM-2007-214996.

orbit. The main reason for increasing the altitude are the constraints imposed by thermal heating

Of the three abort modes, the most desirable is the ATO mode since this allows the possibility of continuing a nominal mission or at least landing the crew within CONUS, thereby ensuring safer recovery operations.

The three abort modes described were subjected to analysis using a software program which breaks down the trajectory sequences into phases, which are then used to model specific parameters of each abort mode. For example, one phase common to each abort mode is the 'abort initiation phase'. Using the software, engineers determine factors such as the earliest possible abort time, given constraints such as vehicle configuration and environment. The software then considers other factors that may affect the abort, such as altitude, latitude and longitude, velocity, relative flight path angle, and azimuth.

The end result of using the software to analyze the three abort modes was to provide engineers with thrust-to-weight ratios, propellant loadings and other factors that may affect the performance of Orion during each abort mode. It also clearly defined abort windows for given engine configurations and confirmed that abort mode overlaps were sufficient and in accordance with Constellation requirements.

Risk assessment

In July 2006, a Systems and Software Consortium (SSC) Risk assessment team generated a risk profile for Orion using a modified Delphi technique developed by the RAND Corporation in the 1940s. The Delphi technique, named after the ancient Greek oracle at Delphi that was believed to predict the future, is characterized by anonymity, controlled feedback and statistical response. To generate the risk assessment, the SSC team followed a three-step process, starting by sending a questionnaire to a panel of experts. Responses from the experts were then iterated and risks categorized, after which a face-to-face meeting with the experts was facilitated, using the Electronic Meeting System (EMS).

Some of the experts questioned were NASA personnel such as Associate Administrator of Safety, Bryan O'Connor, and NASA Risk Analyst, Bill Cirillo. Other experts were non-NASA personnel such as John Kara, Lead of the Advanced Launch Vehicle Program at Lockheed Martin, and ex-NASA veteran, George Mueller, who works as Program Manager for Kistler Aerospace. The experts were sent a questionnaire asking them to provide their considered opinions regarding the risks to Orion's navigation system, ECLSS, propulsion system, and other critical systems. Following iterations of questions and responses among the experts and after conducting the face-to-face meeting via EMS, the SSC risk-assessment team defined nine categories of risks, including risks assigned to systems complexity, systems architecture, re-entry failure modes, and integration, verification, and validation issues. Next, the SSC team formulated fourteen questions addressing each category of risks. For example, Question Six asked what the risks were due to parts, materials, and components selection, while Question Eight asked what risks were associated with the launch. A grid was then generated, tabulating all risks for each question and category, a process resulting in approximately six hundred risks being identified, many of which fell under the categories of system design and development (>150), integration, verification, and validation (>100), and programmatic engineering issues (>90). Some of the major risks to the crew included inadequate TPS and radiation protection, inadequate AR&D, a lack of LAS development, insufficient Orion systems reliability, issues with NASA decision-making culture, and failure of the solid rockets.

Fortunately, since the SSC assessment, many of the risks to Orion have been resolved thanks to a cascade of changes constituting the evolution of the vehicle. For example, mass to orbit problems of the Orion design are slowly evolving away thanks to a redesign of the vehicle and other weight-saving possibilities continue to be studied, such as the module's radiators and changing the TPS material. To give engineers more time to work out some of the weight-saving decisions the Orion program slipped its preliminary design review (PDR) from September to November 2008.

As the centerpiece of the United States' new space exploration program, Orion will have a demanding career. While the vehicle's appearance may seem humble in comparison with its predecessor, the Space Shuttle, the great endeavor that Orion represents is anything but. Although the vehicle's first mission will be to deliver astronauts to the ISS, the focus of Orion's design is ultimately to survive a trip to Mars and back, a mission it will surely achieve thanks to the rigorous design and development process described here.

Regardless of which mission architecture is eventually adopted, Ares V, Ares I and Orion will comprise the integral baseline flight elements supporting the launch of cargo elements and crew into LEO. The final part of this chapter describes the SEI mission assets required to support a manned Mars mission from LEO to Mars and back to Earth. Since such a mission is perhaps two decades in the future, it is expected elements and systems of these assets will be developed and matured.

TRANS-MARS INJECTION STAGE

The TMI stage, inserted into LEO by an Ares V launch, places the ISPS/Return TransHab/ERV stack on a direct Mars-transfer trajectory. The TMI comprises two LOX/LH2 propellant tanks, an intertank adaptor, a payload adaptor, a thrust structure, four RL10B-2 engines and power subsystems.

IN-SPACE PROPULSION STAGE

The ISPS, comprising the return TransHab and ERV (Figure 6.9), provides the MOI burn for the cargo phase of the mission and the TEI for the crewed return phase.

To perform the MOI burn, the ISPS uses an independent set of propellant tanks attached to the core stage of the vehicle. Once the MOI burn is completed, these tanks are jettisoned. The core itself (Table 6.6) comprises two propellant tanks, an intertank adaptor, a payload adaptor, a thrusts structure, three RL10B-2 engines and power subsystems.

IN-SPACE INFLATABLE TRANSFER HABITATS (TRANSHABS)

Role
Separate Inflatable Transfer Habitats (ITHs) are used to support the astronauts during the Earth-Mars and Mars-Earth transit phases. Except for the amount of stored consumables, the TransHabs are identical.

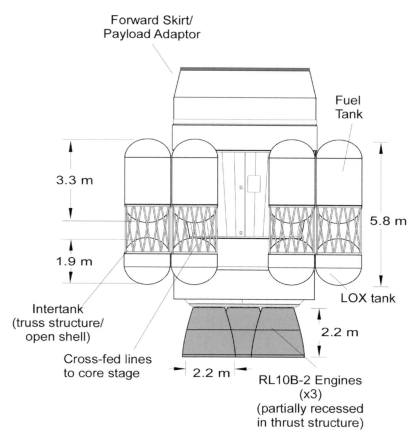

Figure 6.9 SpaceWorks Engineering Inc. in-space propulsion stage. (SpaceWorks Engineering, Inc. (SEI).)

Table 6.6 In-space propulsion stage

Element	Value
Total stage length (including deployed engines)	12.8 metres
Maximum stage diameter	9.2 metres
Core diameter	5.5 metres
Propellant load core stage tanks	2.5 tonnes
Propellant load, fuel and oxidizers	14.8 tonnes
Drop-tank propellant load	4.5 tonnes
Drop-tank oxidizer propellant load	26.2 tonnes

Construction

The TransHab provides forty-three cubic meters of habitable volume and is constructed of a multi-layer, 30-centimeter thick, inflatable shell made of extremely high-strength lightweight fibers. The outer layers of the shell, designed to fragment space debris such as micrometeorites that may strike the shell at speeds as high as seven kilometers per second, protect the inner bladders of the shell, which not only contain the habitat's air but also insulate the habitat from the extreme temperatures of deep space. The exterior layers protect the habitat using alternating layers of Nextel and several layers of foam. Any particle impacting the Nextel/foam combination shatters, losing progressively more energy as it continues to penetrate. Even if the particle manages to punch through the Nextel and foam, it will likely be stopped by a layer of bullet-proof Kevlar surrounding the three air-tight bladders of the habitat.

Systems

Power is provided by four solar arrays, and thermal control is supplied by a two-fluid water/Freon active system that works by pumping water through the central core of the habitat and through coldplates attached to all the major equipment. As the water is being pumped, it carries away heat to the upper and lower parts of the habitat, where the heat is transferred to the Freon loop via heat exchangers. Freon is then pumped through radiators exposed to the vacuum of space, where the heat is finally rejected.

Life support

The closed water and oxygen LSS filters water using multifiltration[1] and reclaims wastewater by vapor compression distillation, a process that boils the wastewater to produce and collect water that is almost completely free of minerals, chemicals and microbes. Carbon dioxide generated by the crew is combined with tanked and generated hydrogen to produce methane and water in a Sabatier reactor (see Glossary). The methane is vented to space and the water is converted to oxygen and hydrogen via electrolysis in an oxygen generation unit.

MARS EXCURSION VEHICLE

Role

In SEI's architecture, both MEVs land on the surface of Mars. Each MEV (Table 6.7) comprises a propulsive descent stage and a payload housed within an outer heatshield. In addition to these elements, the crewed MEV also includes an ascent stage and a pressurized rover, whereas the surface MEV contains the fully-provisioned surface habitat and nuclear surface power system (Figure 6.10).

---

1 Multifiltration removes particulate material by filtration, and dissolved salts and organics are removed by adsorption onto sorbent media. Any material that is not effectively removed by sorption is destroyed by catalytic oxidation.

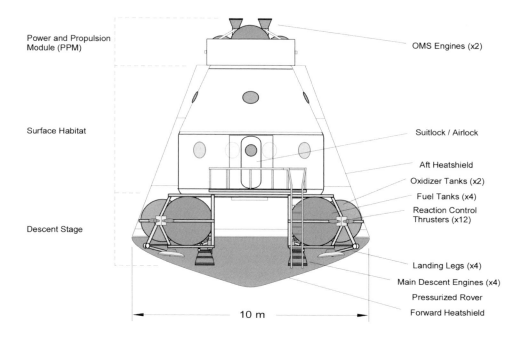

Figure 6.10 SpaceWorks Engineering Inc. Mars Excursion Vehicle. (SpaceWorks Engineering, Inc. (SEI).)

Table 6.7 Mars Excursion Vehicle, integrated configuration

Surface Habitat		Crewed MEV	
Maximum Diameter	**10 m**	Maximum Diameter	**10 m**
Surface Habitat	**20.3 t**	Pressurized Rover	**4.1 t**
Power & Propulsion Module	**2.9 t**	Ascent Stage	**14.4 t**
Descent Stage	**8.4 t**	Descent Stage	**9.9 t**
Forward Heatshield	**2.1 t**	Forward Heatshield	**2.1 t**
Total MEV Mass	**33.7 t**	Total MEV Mass	**30.5 t**

Heatshield

In common with the Orion TPS, the MEV's heatshield is made of PICA and includes three component layers which are, from outermost to innermost, PICA, a metallic honeycomb support structure and an insulating blanket material. The heatshield diameter is ten meters, a dimension baselined in order to provide sufficient aerocapture performance while still maintaining compatibility with the Ares V shroud dimensions. However, computer modeling suggested a ten-meter diameter heatshield will not provide sufficient drag to slow the vehicle into LMO, even when combined with other aerocapture systems, so it is probable the final heatshield design will be a collapsible version that extends to a greater diameter.

Descent stage
The MEV descent stages are sized to provide for slightly different propellant loads but this does not result in many structural differences. Based on an open truss structure design the descent stage houses four rocket engines and spherical propellant tanks containing liquid oxygen and liquid hydrogen. Also enclosed within the truss structure are various subsystems, such as the battery bank providing power for the surface habitat MEV descent stage during the EDL phase.

Ascent stage
Used to transport the crew and a payload from the Martian surface to LMO, the ascent stage comprises a pressurized cylinder housing the crew, four main rocket engines, two orbital maneuvering system (OMS) engines and an array of reaction control thrusters. To provide access to the TransHab and the pressurized rover, the pressurized cylinder is fitted with two hatches, one on the upper and one on the lower part. The LOX and LH2 propellants are stored in six spherical tanks located around the crew cabin. Electrical power is provided by an Advanced Stirling Radioisotope Generator (ASRG).

MARS SURFACE HABITAT

The two-level MSH (Figure 6.11) provides eighty-four cubic meters of habitable volume and supports the three crewmembers during their stay on Mars. Access is via a hatch in the floor connecting to the pressurized rover, and two suit airlocks enable crew extravehicular activity (EVA) egress. During the day, the crew will spend most of their time on the lower level which houses the galley, an airlock, a small science laboratory, a medical station and a hygiene area. The upper level is split into crew quarters. Resistive exercise equipment is conspicuously absent since the crew will spend a significant part of their time performing EVA operations. Another design item not featured in the MSH is a radiation shelter, since it was argued the Martian atmosphere provides sufficient protection from solar particle events (SPEs) [5]. However, this feature may change depending on data from future robotic missions designed to characterize the Martian radiation environment.

Life support system
The MSH Environmental Control and Life Support System (ECLSS) employs a vapor phase catalytic ammonia removal water recovery system. Food is supplied as prepackaged individual entrees but, due to the mission duration, the MSH includes a biomass production chamber providing the crew with the opportunity to supplement their meals with various salad items. Waste is stored without claiming any products from the waste. To facilitate cooling of the vehicle, coldplates and condensing heat exchangers are used to collect heat. Clothing is used once and then discarded to the waste subsystem after one week of use. Bacteria is filtered from the atmosphere using a filter assembly consisting of a fiberglass and aluminum housing assembly containing a replaceable high efficiency particulate air (HEPA) filter capable of

Figure 6.11 SpaceWorks Engineering Inc. Mars Surface Habitat. (SpaceWorks Engineering, Inc. (SEI).)

filtering 99.9% of all particles 0.3 micron or larger. A Common Cabin Air Assembly (CCAA) provides temperature and humidity control. To remove all the heat from the racks of equipment in the MSH, an avionics air assembly (AAA) is fitted, consisting of a compact heat exchanger and an extremely high-speed fan housed in a tightly-packed assembly.

Air

A regenerative carbon dioxide removal system, based on molecular sieve technology, is used to absorb carbon dioxide which is dumped overboard and not recovered. A trace gas contaminant system removes trace combustible gases from the MSH environment, and a bacteria filter assemblies filter out any particulate matter. Oxygen is supplied from high-pressure stores and also as a product from the electrolysis of water, using solid polymer technology. Detection of air contaminants is provided by the constituent analyzer forming a part of the fire detection and suppression system.

Biomass

The biomass production system (BPS) provides the crew with salad crops as a welcome supplement to the prepackaged food system. Although the dietary nutrients provided by salad crops are negligible, their availability to the crew provides a psychological benefit that will become more pronounced as the mission progresses. The biomass chamber comprises four plant growth chambers, each providing a third of a square meter of growth area. Although the BPS is fully automated, the crew conducts periodic system checks and support tasks such as refilling water reservoirs and changing carbon dioxide canisters.

Food

The MSH is stocked with a variety of fresh, dehydrated, water-preserved, shelf-stable and frozen foods. In addition to individual entrees, crewmembers supplement their diet with fresh salad from the BPS and snacks.

Thermal

The MSH is equipped with two thermal management systems. The internal thermal control system includes the avionics air assemblies, ensuring air-cooling for equipment, and the cabin air assemblies which cool, dehumidify and circulate cabin air. The external thermal control system includes the radiators and heat exchangers that transfer heat from the inside of the vehicle to space.

Waste

Trash, fecal material and all forms of solid waste are stored onboard the MSH.

Water

The price tag of a glass of water on the ISS is about three thousand dollars, thanks to the exorbitant costs of ferrying cargo into LEO. However, NASA has said enough is enough and, from 2009, astronauts will be drinking their own urine from a water recovery system that the space agency has spent decades perfecting. The Russians developed a similar system in the 1980s but it never flew in space because of concerns over crew squeamishness. However, expedition crews can't afford to be squeamish, and NASA can't afford to foot the bill of providing all the water required for a Mars mission, so the MSH will be fitted with a similar water recovery system to the one that will be fitted on the ISS.

   The process of water of recovery is very simple. The water recovery system (WRS) collects urine from the astronauts and condensation from the cabin air and, via a series of chemical treatments and filters, turns the moisture into drinkable water. The first part of the process involves filtering out all the solid particles, such as skin cells, out of the liquid. Once contaminants are removed, they are chemically dissolved and oxygen is added to the liquid to oxidize any trace material, so they too can be removed. Any chemicals left over from the cleaning process are removed and, finally, iodine is added to reduce microbial growth and the liquid is sent to the storage tank ready for drinking.

   Unsurprisingly, the water from the tanks will not taste like regular tap water.

NASA has conducted blind taste tests comparing recycled drinking water with plain tap water as well as with tap water with iodine added. On a one to nine scale, none of the waters rated higher than five. However, while the taste may not be very appealing, perhaps the most unsettling aspect for the astronauts drinking the water is not knowing whose urine they are drinking!

Extravehicular activity support
The basic extravehicular mobility unit (EMU) consumables include water, used for thermal management by means of an internal cooling garment, and oxygen, used for metabolic consumption and suit pressurization. The basic EMU uses lithium hydroxide for carbon dioxide removal. Egress is via the MSH's airlock, which uses an airlock pump to reduce the airlock internal pressure to about ten percent of the cabin pressure. A High Pressure Oxygen Generator Assembly (HPOGA) is used to produce high pressure oxygen to recharge the EMU.

PRESSURIZED ROVER

Included in the MEVs payload is a four-wheeled three-person pressurized rover, first used after landing to transport the astronauts to the MSH, and, at the end of surface operations, to transport them back to the crewed MEV. During surface operations, the ASRG-powered rover is used for excursions in support of surface operations in the local and regional area.

IN SUMMARY

This chapter has provided an overview of the hardware supporting the SEI Mars architecture. While a manned Mars mission may be two decades in the future, it is important, even at this early stage, to evaluate how flight assets and hardware generated by program such as Constellation may translate into future missions, such as the one proposed by SEI. Equally important is an assessment of how crews may be selected for Expedition Class missions and how they may be trained. These issues are discussed in the following chapter.

REFERENCES

1.   Advanced Transportation System Studies Technical Area 2 (TA-2). Heavy Lift Launch Vehicle Development Contract. NAS8-39208. DR 4. Final Report. Prepared by Lockheed Martin Missiles & Space for the Launch Systems Concepts Office of the George C. Marshall Space Flight Center. (July 1995).
2.   Constellation Architecture Requirements Document. NASA CxP 70000. Baseline. December 21, 2006.

3.   Exploration Systems Architecture Study. Final Report. *NASA-TM-2005-214062.* (November 2005).
4.   NASA. *Man-Systems Integration Standards. NASA STD-3000.* Houston, TX: NASA Johnson Space Center. 1995.
5.   Simonsen, L.C., and Nealy, J.E. Mars Surface Radiation Exposure for Solar Maximum Conditions and 1989 Solar Proton Events. NASA TP-3300, 1993.
6.   St. Germain, B.; Olds, J.R.; Bradford, J.; Charania, A.C.; DePasquale, D.; Schaffer, M.; Wallace, J. Utilizing Lunar Architecture Transportation Elements for Mars Exploration. SpaceWorks Engineering, Inc. (SEI), Atlanta, GA, 33075. 2007.

7

Crew selection and training

"Men wanted for hazardous journey. Low wages, bitter cold, long hours of complete darkness. Safe return doubtful. Honour and recognition in event of success."
   Advertisement rumored to have been posted by Sir Ernest Shackleton before the launch of his legendary 1914 Imperial Trans-Antarctic Expedition.

The advertisement allegedly placed by the great Anglo-Irish explorer Ernest Shackleton may be apocryphal, but its content applies equally to those selected for a Mars mission, an expedition characterized by extreme temperatures, galactic cosmic radiation, high-speed micrometeorites and a host of physiological and environmental stressors. The expeditions embarked upon by Shackleton, Fridtjof Nansen and Douglas Mawson almost a century ago resemble in many ways the conditions of isolation and confinement which will be experienced by future space travelers traveling to and living on Mars. The conditions will be different, but many of the problems confronting future space explorers will be the same ones that troubled explorers in the past, a reality that will be reflected in the unique selection criteria applied to those lucky few chosen to go to Mars.

   This chapter describes how Expedition Class crews may be selected and trained, how space agencies will decide who has the 'right stuff' for a long duration mission and what factors, beyond technical skills and education, should be considered for selecting a crew for what will be the most arduous human expedition to date.

CREW SELECTION

"The human factor is three quarters of any expedition."
   Legendary Norwegian polar explorer, Roald Amundsen.

In common with the crewmembers of Nansen's and Shackleton's expeditions, the austere and isolated conditions facing future Mars explorers will impose significant hardship upon those selected. While it will be assumed an astronaut has the skills and knowledge necessary to perform the duties of a crewmember, these abilities will count for nothing if he or she cannot get along with others for four months in the confines of a vehicle the size of school bus! Given the unique characteristics of such a mission, the issues of crew composition and crew compatibility clearly become factors almost as important as the selection process itself. Due to the potential for these issues to impact negatively upon a mission, it is important to address them before moving on to discussing the selection process itself.

Crew composition

Crew Size
The European Space Agency (ESA) plans to send a crew of four to the Red Planet, NASA is preparing to send six astronauts, and the SpaceWorks Engineering Inc. (SEI) architecture is designed to transport just three crewmembers. Regardless of which mission plan is ultimately adopted, it is likely the size of the crew will fit with the current belief that smaller is better. Such a policy was implemented on many of the most successful polar expeditions such as Nansen's three year *Fram* venture which comprised just thirteen crewmembers and Shackleton's Imperial Antarctic Expedition, which consisted of just twenty-seven. Despite extreme isolation and prolonged confinement, Nansen's and Shackleton's expeditions were characterized by few interpersonal problems, thanks largely to the small homogeneous crews, a lesson unlikely to be overlooked when it comes to defining the composition of a Mars crew.

Crew roles
The occupational role of each member of the Martian crew has yet to be determined, but it is almost certain one crewmember will be a pilot and it is likely, given the extended duration of the mission, another crewmember will be a medical doctor,. The role of commander will be assigned to the crewmember with the most experience and will not necessarily be the pilot, as has been the case in so many space missions. Given the science objectives of such a mission, it is inevitable the crew will include at least one scientist, and other crewmembers will be cross-trained in various scientific disciplines.

Crew gender
The issue of whether a crew should be all-male, all-female or mixed remains a contentious matter. Some have argued a female crew would exhibit preferable interpersonal dynamics and be more likely to choose non-confrontational approaches to solve interpersonal problems, whereas others have made a case for a mixed crew, claiming crews with women are characterized by less competition and seem to get along better. Evidence from Antarctic winter-over crews supports each of these arguments and suggests the inclusion of women in space crews would serve a socializing purpose, in addition to their mission function. However, the introduction of a single female into a male group may have destabilizing effects because of sex issues, a topic that space agencies are reluctant to discuss. What effect would a passionate affair during a mission to Mars, for example, have upon other

crewmembers and on overall crew performance? Fortunately, for mission planners at least, research indicates a drop in the sex hormone levels of astronauts during long-term missions, resulting in a pronounced decline in sex drive, so maybe the sex problem won't be an issue after all. However, just to be on the safe side, it is likely the first mission will comprise an all-male crew. Such a decision is based not only on the psychological issues but also because males are more resistant to radiation, thereby reducing the overall mission risk.

Crew compatibility
Determining crew compatibility has often been viewed as an opaque process. Perhaps the most diverse crew ever launched into orbit was Space Shuttle mission, STS-51G, comprising civilian and military NASA astronauts, both male and female, a Saudi Arabian prince, and a French cosmonaut. Despite the obvious multinational nature of the mission and the obvious cross-cultural challenges, the crew of STS-51G was, by all accounts, a harmonious one that was compatible and worked effectively throughout the mission. However, typical Space Shuttle missions last no more than two weeks, whereas an Exploration Class mission will last two years or more. Crew compatibility issues will therefore obviously assume increasing significance in determining the effectiveness of the mission.

The problem in determining crew compatibility is there is no one measure to predict whether a crew will work together effectively. Some researchers favor the use of psychological performance tests and personality questionnaires, whereas other investigators prefer a more behavior-oriented approach. The Russians, on the other hand, who have invested considerably in developing methods to assess interpersonal compatibility, consider biorhythms to be a useful tool for selecting cosmonauts. Once again, useful lessons can be found in the annals of polar exploration and research conducted during Antarctic winter-over increments which suggest personality traits and interpersonal skills be carefully scrutinized when selecting crews for long-duration missions. However, all the psychological tests in the world will be unable to predict how crewmembers will interact. To resolve this, a candidate crew will probably spend several weeks in a high-fidelity simulation to demonstrate to themselves and mission managers they can adapt to the many unique stressors associated with living and working in close proximity.

A veritable cornucopia of knowledge regarding crew composition and compatibility exists thanks to the successful expeditions of Shackleton, Nansen and Amundsen and experience from Antarctic research stations. This history of crew dynamics in harsh environments provides mission planners with more than enough information to carefully select a compatible crew. But what does the actual selection process entail?

Crew selection overview
Assuming the first Mars crew departs Earth in February 2031, a call for selection will be advertised in 2026, allowing sufficient time for the selection process, astronaut training, mission preparation for the crew's first mission, and Mars-mission specific training. Crew selection procedure will follow NASA's current basic health and

Table 7.1 Crew selection requirements

Personal Requirements	
Male	Age: > 50
Meet space agency medical standards	Possess sense of community
Free of psychological problems	Possess effective conflict resolution skills
Technically Competent	Possess sense of teamwork.

Medical Requirements	
Genetically screened for disease.	Above average bone density.
Undergone appendectomy.	Genetically screened for radiation resistance.
Undergone gall bladder removal.	Screened for kidney stones.

Social Skills and Behavioural Traits	
Social Compatibility	Tolerance
Emotional Control	Agreeable and flexible
Patience	Practical and hard-working
Introverted but socially adept	Does not become bored easily
Sensitive to the needs of others	Desire for optimistic friends
High tolerance for lack of achievement	High tolerance for little mental stimulation
Self-confident without being egotistical.	Subordination of own interests to team goals.

Crew Compatibility Traits	
Tactfulness in interpersonal relations	Effective conflict resolution skills
Sense of humor	Ability to be easily entertained

education selection requirements, with all crewmembers having either a science or engineering background. Crewmembers will also be required to meet a number of unique Expedition Class mission requirements, some of which are described here and summarized in Table 7.1.

Selection criteria unique to Mars missions
Due to the unique characteristics of such an extreme mission, space agencies will need to employ some rather unconventional selection criteria such as genetic screening and precautionary surgery. Given the unusual nature of these criteria, it will require those selected to cross a legally defined boundary, in the same way a soldier does when joining the military, by relinquishing certain individual rights and accepting collective standards contributing to the common good of realizing a successful mission.

Genetic screening
Although current United States legislation [7] bars employers from using the genetic information of individuals when making hiring decisions, NASA will surely be exempted when it comes to recruiting astronauts for Mars missions.

Rationale

It is likely potential astronaut candidates (ascans) will be genetically tested as a part of the medical during the astronaut recruitment campaign. Genetic testing allows the space agencies to diagnose vulnerabilities to inherited diseases and also reveals information concerning the presence of genetic diseases and mutant forms of genes associated with increased risk of developing genetic disorders. Furthermore, genetic testing can confirm or deny a suspected genetic condition and provide information concerning the possibility a person may develop a disorder.

Types of testing

Diagnostic testing will be used to eliminate most genetic or chromosomal conditions, while carrier testing will be used to identify candidates carrying one copy of a gene mutation that may cause a genetic disorder. Predictive testing will be used to detect gene mutations associated with disorders present in the candidate but where no features are present at the time of testing. This type of testing will identify candidates at risk of developing a disease such as cancer during a mission. Obviously, if any of the results are positive, the candidate will be eliminated from the recruitment process.

The tests described are performed on a sample of blood, hair or skin, which is sent to a laboratory where technicians search for differences in chromosomes, DNA, or proteins. Due to the problems in interpreting genetic tests, space agencies will need to exercise particular care in determining the genetic profile of candidates. For example, a negative test result means the laboratory did not detect an abnormal gene, chromosome or protein, but, although such a result may indicate a person is not affected by a particular disorder, it is possible the test missed a disease-causing genetic alteration because some tests simply cannot detect all genetic changes associated with a specific disorder. To eliminate any ambiguity, space agencies will undoubtedly discard uninformative and inconclusive tests and conduct secondary tests but, in the event of a positive result, the likely consequence for the candidate will be elimination from consideration as an astronaut. Needless to say, the effect of a positive result on candidates who have spent their entire professional lives accumulating the qualifications to become an astronaut will be devastating. However, such testing is a necessity, given the potential consequences of an astronaut being diagnosed with a critical illness during the mission.

Precautionary surgery

Once a crewmember is pronounced genetically free of any future disease or disorder and is provisionally selected for the crew of the Mars mission, they will probably be required to undergo precautionary surgery. While a candidate may be required to undergo a number of precautionary procedures, the most likely one to be performed is removal of the appendix.

The appendix

The appendix is a closed-end narrow tube attached to the first part of the colon. If the opening to the appendix becomes blocked or the fatty tissue in the appendix swells, bacteria, which are normally found within the appendix, may invade and infect the wall of the appendix, resulting in *appendicitis*. The body responds to this bacterial attack by inflaming the appendix, which may ultimately lead to rupture, followed by spread of bacteria outside the appendix. Alternatively, the appendix may become perforated which may lead to an abscess or, in some cases, the entire lining of the stomach may be infected.

Appendicitis
Less common complications of appendicitis include blockage of the intestine, a condition in which the intestinal contents are prevented from passing, resulting in distension of the stomach. Such a condition may require the contents of the stomach to be drained through a tube passed through the nose. Needless to say, in the confined environment of the crew vehicle exposed to zero gravity, such a procedure would challenge even the most experienced surgeon!

Sepsis
Perhaps the most feared complication of appendicitis is *sepsis*, a condition in which bacteria enter the blood and infect other parts of the body. Even on Earth, sepsis is considered a serious complication, but to an astronaut bound for Mars or returning to Earth, such a complication would be a death sentence.

Medical support
While these complications alone represent a powerful argument for removing the appendices of Mars-bound crewmembers, there are also other factors to consider such as the diagnosing of the condition, a procedure which would use vital medical consumables. For a crewmember suspected of suffering from appendicitis, the only possible diagnostic procedures available would be a urinalysis and an ultrasound procedure. In the event of complications, a computer tomography (CT) scan and abdominal X-ray would be unavailable due to the limited medical resources onboard, although it is possible a laparoscopy could be performed. However, laparoscopy, a procedure in which a small fibreoptic tube with a camera is inserted into the abdomen through a puncture made in the wall of the stomach, requires a general anesthetic and would present a challenging procedure in zero gravity. Furthermore, even on Earth, appendicitis is often difficult to diagnose because other inflammatory problems can mimic the symptoms of the condition.

Appendectomy
Should a crewmember be correctly diagnosed with appendicitis the next problem is treatment, involving removal of the appendix in a procedure known as an *appendectomy*, which requires the surgeon to make a four to six centimeter incision through the skin and layers of the abdominal wall, in the area of the appendix, and remove the appendix. If an abscess is present, pus must be drained before the abdominal incision is closed. In recent years, laparoscopic surgery is being used to perform the procedure but, in zero gravity, either method would present risks.

   Once the candidate Mars crewmember has successfully navigated the challenging

Table 7.2 Crew training timeline for first flight

Ascan		Astronaut							
BASIC TRAINING	Certification	SYSTEM-RELATED TRAINING	Certification	REFRESHER TRAINING	Certification	MISSION-RELATED TRAINING	Certification	FLIGHT (Lunar)
12 months		9 months		Ongoing		18 months		2 weeks

selection process, been genetically screened, and has undergone the requisite precautionary surgery, they will undoubtedly breathe a sigh of relief and look forward to the training for the mission!

CREW TRAINING

Basic crew training

The newly selected astronaut candidates (ascans), will spend twelve months in basic training. During this time they will learn essential operational knowledge of the Ares I and Ares V launch vehicles and the Orion crew vehicle. Basic training will also include courses in scuba-diving, survival training, parachuting, and aircraft safety in addition to the ongoing training in piloting and language skills, and science and technical instruction (Table 7.2).

Once acquainted with the Ares V, Ares I and Orion, ascans will spend time becoming familiar with Mars flight elements through lectures, briefings, textbooks, mockups, and flight operations manuals. To help familiarize themselves with the myriad systems and subsystems of the flight elements, ascans will use the single system trainer (SST) which will teach them how to react in malfunction and contingency situations. Another important feature of basic ascan training will be microgravity training, which will take place in the Weightless Environment Training Facility (WET-F) and onboard the specially modified aircraft (Figure 7.1).

On completion of basic training, ascans will undergo a certification process including written exams, simulated tests, interviews, and a review by board. On successful completion of certification, ascans become astronauts but are not eligible for a flight assignment until one year after completion of basic training. This is due to the additional training requirements for their pre-Mars expedition which will be a two-week mini-Mars mission to the Moon.

By 2026, a lunar outpost will have been established and since aspects of the lunar mission architecture are similar to the Mars mission, a trip to the Moon will provide the crew with experience in the procedures they will perform during their trip to Mars. For example, they will be able to practice rendezvous and docking in low Earth orbit (LEO), low lunar orbit (LLO), descent and landing procedures, ascent from the lunar surface and of course, re-entry procedures. Due to the impact of

Figure 7.1 The Airbus A300 performing parabolic flight. (Novespace.) See Plate 11 in color section.

radiation and bone demineralization on the crew during a Mars mission, the lunar trip will be the only space mission the Mars crew will fly before embarking for Mars but, before describing the warm-up mission to the Moon, let's look at the crew's mission training.

Pre-Mars mission-related training

Following completion of basic training, the crew prepares for their two-week warm-up mission to the Moon. The commander and pilot conduct identical training, so the pilot is capable of taking over the commander's role and vice versa. To ensure an additional level of redundancy, a crewmember is also fully trained in piloting duties. During the mission-related training phase, a crewmember is designated Crew Medical Officer (CMO) and crewmembers immerse themselves in system-related training and vehicle familiarization training, and with the lunar lander and outpost systems. To fine-tune their skills, crewmembers spend time in medium-fidelity trainers, becoming familiar with single and multi-system operations in nominal and off-nominal modes. The commander and pilot conduct additional training in the Lunar Landing Training Vehicle (LLTV), configured to simulate the handling characteristics of the lunar lander. Towards the end of the mission-related training phase, the crew spends more and more time practicing entire phases of the mission and flight procedures in full systems mockups to ensure they learn the corrective actions to single-system and multi-system failure modes. Finally, on completion of the mission-related training phase, the crew undergoes a certification process comprising exams, simulated tests, interviews and a review board.

Pre-Mars mission

The trained Mars crew now embarks on its first flight in which they experience most of the elements of the Mars mission architecture. The mission commences with the launch of Ares V, placing the Lunar Surface Access Module (LSAM) and Earth Departure Stage (EDS) in LEO. The crew launches in Ares I, placing Orion in LEO, whereupon Orion and the LSAM/EDS configuration rendezvous and dock. The EDS fires its two J-2S LOX/hydrogen engines, sending the stack on a trans-lunar trajectory (TLI), after which the EDS is expended. The LSAM and Orion fly to the Moon and are inserted into lunar orbit. Once stable in lunar orbit, the crew descends to the surface in the two-stage lunar lander, comprising a descent and ascent stage, leaving the unmanned Orion behind in orbit where it will operate autonomously during the crew's seven day surface stay. After its lunar stay, the crew fires up the ascent stage's pressure-fed LOX/methane engines and returns to lunar orbit, docks with Orion, transfers back into Orion and departs lunar orbit, using the SM propulsion system of the CEV. Finally, Orion performs a direct-Earth-entry followed by a parachute water/land landing.

Mars Mission training

Shortly after returning from the Moon, the six crewmembers are confirmed as the first Mars crew, a decision initiating a lengthy training phase, culminating in the first manned flight to Mars. Although the program will follow a similar sequence to the lunar mission-related training, the preparation will require two or more years (Table 7.3) due to the requirement of becoming familiar with all the flight elements and also analog and surface operations training.

Table 7.3 Crew training timeline for Mars mission

SYSTEM RELATED TRAINING	Certification	REFRESHER TRAINING	Certification	MARS MISSION TRAINING	Certification	FLIGHT TO MARS
18 months		Ongoing		24 months		30 months

Emergency training

During their training, astronauts will become familiar with JSC's Building 9 which houses a practice arena in which astronauts perform emergency scenario training. The emergency scenarios will prepare astronauts to respond to fire/smoke, rapid/explosive decompression and toxic release events. In each of the scenarios, time will be critical. For example, a micrometeorite punching a one-centimeter hole in the surface habitat may give astronauts only minutes to respond before their air supply is vented to space as a result of rapid decompression of the habitat.

Psychological training

Once a Mars crew has been selected, one aspect of their training will be directed at developing interpersonal dynamics not only between each other but also between the crew and ground control personnel. This type of training is already common in both the American and Russian space programs and it will become even more important as missions become longer and crews are exposed to the dangers of radiation and microgravity for extended periods.

To help develop their interpersonal dynamics, astronauts will be assisted by the NASA Psychological Services Group (PSG) composed of a group of behavioral scientists and psychologists. It was designed to support American astronauts during their stays onboard Mir and will no doubt play an equally important role for astronauts embarking upon a six-month tour of duty on the surface of Mars. In addition to helping astronauts adapt psychologically, the PSG will provide input on a number of mission variables, ranging from work and rest schedules to training crewmembers to recognize potential in-flight psycho-sociological issues.

Another important support unit tasked to help the astronauts and their families will be the Family Support Office (FSO), a group including representatives from the PSG, the Astronaut Office, and the Astronaut Spouses Group (ASG). While astronauts are deployed to Mars, the FSO will serve as a liaison with NASA, providing regular contact with family members and updating information regarding the mission. For example, some of the events the FSO may be involved with include organizing family conferences on special occasions, compiling digital family picture albums, and debriefing the family following the mission.

Virtual environment generator training

Astronauts will also be trained in the use of the Virtual Environment Generator (VEG). The VEG [11, 13] is a virtual reality (VR) system that can simulate certain aspects of microgravity, assist in navigating new environments, such as the Mars habitat, and serve as a countermeasure to spatial disorientation. The VEG comprises a head-mounted display (HMD), the position and orientation of which commands a computer to generate a scene corresponding to the position and orientation of the operator's head [3]. This synthetic presence permits the operator to move around in the artificial world of the Mars habitat or even traverse the Martian surface.

When astronauts don the VEG equipment, they will be presented with an image of the interior of the habitat and a space-stabilized virtual control panel with an image of the astronaut's hand in the HMD. As the astronauts move their hands, the virtual hand will also move. Collision detection software in the graphics computer [4, 6, 12] detects when the operator's hand penetrates the virtual control panel, enabling the astronaut to interact with the virtual switches or objects to control events within the habitat.

Astronauts will also be able to manipulate objects in the virtual habitat and be able to experience resistance to movement, texture, mass and compressibility, thanks to the haptic (tactile) and force feedback [4, 14] systems. To help astronauts in the virtual habitat, the system has been designed to provide auditory cues when an object is grasped or dropped, or when a virtual switch is operated. This synthesis of visual and auditory cues will augment the visual information presented to the operator, thereby enhancing the performance of crewmembers within the habitat.

Data compression techniques result in the virtual habitat containing all objects one would expect to see in the real habitat. Software also takes into account the effect of human behavior and the effect of collision for real-time operation, which means no matter how fast the operator moves through the environment, he/she experiences no visual lags [12]. The real-time operation results in the operator experiencing the high degree of realism and interactivity necessary to allow crewmembers to perform tasks necessary for training.

Cryopreservation indoctrination
It is known short-duration spaceflight has no adverse effect on the ability of astronauts to conceive and bear healthy children to term. However, pregnant astronauts are not allowed to train in the vacuum chamber, KC-135 Zero-G aircraft, fly in T-38s or train in the Neutral Buoyancy Facility (NBF). Due to the training requirements for a Mars mission, any astronaut becoming pregnant[1] will likely be dropped from their flight assignment. Another problem for astronauts who wish to conceive is the effect of a long duration mission upon fertility and the increased chance of genetic defects as a result from prolonged exposure to deep space radiation. Since the effects of a return trip to Mars upon the reproductive capacity of humans have yet to be characterized, it is difficult to provide guidance regarding the risks of radiation damage, so the obvious solution will be to provide a program for both male and female astronauts to cryopreserve embryos on Earth for future use. For women, *banking* will eliminate the potential problems of damage to embryos caused by galactic cosmic radiation and solar particle events and also augment fertility since the pregnancy and miscarriage rates for embryo transfer are dependent on the age of the embryos at the time of collection. Another option open to female astronauts will be to cryopreserve ovarian tissue. For male astronauts, sperm cryopreservation will be implemented as an option in case of their returning to Earth infertile.

Hibernation familiarization
Another important element of training will be a theoretical and practical familiarization with the process of hibernation. While this procedure is presently still in the realm of science fiction, it is possible the technology may be operational by the time the first crew departs in the 2031 to 2035 timeframe.

---

1 Although it is proposed the first mission be all male, mixed crews will follow once more radiation resistant materials have been developed.

Table 7.4 Effect of hibernation on life support requirements [1, 2]

Life Support Component	Purpose	Effect of Hibernation
Atmosphere Management	Air revitalization, temperature, humidity and pressure control Atmosphere regeneration. Contamination control	Reduced heating requirement. Reduced regeneration requirement
Water Management	Provision of potable and hygienic water. Recovery and processing of waste water	Reduced significantly
Food Storage	Provision of food	Reduced significantly
Waste Management	Collection, storage and processing of human waste	Reduced significantly
Crew Safety	Fire detection and suppression Radiation warning system	Augmented systems required
Crew Psychology	Maintenance of crew mental health	Reduced
Crew Health	Bone demineralization and muscle atrophy	Augmented systems required

If thoughts of long duration space journeys and hibernation conjure up images of the opening scene of *Alien,* you're not alone but this technology no longer exists merely in the realm of science fiction movies as both NASA and ESA are funding research into methods permitting astronauts to spend months in a state of suspended animation. Although the concept may seem more fiction than science, the daunting timeframe facing astronauts means hibernation is an idea to be taken seriously. Apart from the boredom of a lengthy transit, there are powerful logistical reasons to place astronauts in hibernation. Both ESA and NASA have estimated a typical two-year return trip to Mars would require thirty tonnes of consumables for a crew of six. In addition to the food issue, there are the not inconsiderable matters of waste generation and oxygen consumption. Hibernating astronauts require less oxygen and food than active astronauts, thereby resulting in a lighter spacecraft, less fuel and less fuel (Table 7.4)! Already, scientists and engineers are designing 'sleep pods' that may resemble those on the Nostromo, the spacecraft in the film *Alien.*

Of course the next question is whether hibernation will actually work. Well, the scientists working for ESA's Advanced Concepts Team (ACT) seem to think so and have studied other species, such as bears, ground squirrels and rodents, for which hibernation is a regular part of their lives. Already researchers have been able to chemically induce a stasis-like state in living cells and have progressed to small, non-hibernating mammals like squirrels (Figure 7.2).

Figure 7.2 California ground squirrel (*Spermophilus beecheyi*). (Free Software Foundation/ Wikipedia.)

The key to putting astronauts in a state of hibernation may lie in a synthetic, opioid-like compound called *Dadle*, or *Ala-(D) Leuenkephalin*, which, when injected into squirrels, can put them in a state of hibernation during the summer [2]. This research has been already extended to studies investigating the effect of applying Dadle to cultures of human cells, an investigation revealing human cells divide more slowly when Dadle is applied. In conjunction with the studies investigating Dadle, researchers are testing compounds such as *dobutamine* and *insulin-growth factor* (IGF). Dobutamine is normally administered to bedridden patients to strengthen heart muscles but, in the case of hibernating astronauts, the compound would be administered to maintain health during the long period of inactivity. Insulin-growth factor would be administered to boost the astronauts' immune systems, which would be depressed during the long period of inactivity.

How would the system work? First, crewmembers would be required to attain a very high level of fitness to maximize their body's ability to deal with the stress of hibernating and due to the unavoidable deleterious effects of hibernation. Shortly after performing trans-Mars insertion (TMI), astronauts would place themselves in the Mars Transit Vehicle's *hibernaculum* and connect one another to intravenous tubes, through which fluids and electrolytes would be administered to compensate for changes in blood composition during the hibernation. Then, administration of a hibernation-inducing compound would place the astronauts in a state of hibernation [1]. During the hibernation, a suite of medical sensing and hibernation administration facilities would monitor the state of the hibernating astronauts. In addition to ensuring body temperature, heart rate, brain activity and respiration stay within normal boundaries, the medical equipment would also monitor blood pressure, blood glucose levels, and blood gases. In addition to the medical equipment, medical monitoring would also be conducted from Earth, although the time delay would limit the ability of mission control to react in the event of an emergency.

To guard against contingencies, the hibernation suite would be equipped with in-situ monitoring by an artificial intelligence (AI) agent operating on principles similar to medical monitoring systems such as GUARDIAN [9]. The agent would monitor the hibernation medical equipment, ensure environmental parameters are maintained and monitor contingency events that may require waking of the crew such as a solar storm or a mid-course correction.

Although placing astronauts into hibernation would solve many problems during the deep space phases of a Mars mission, several issues still must be resolved. Scientists still need to develop a trigger compound capable of inducing a state of hibernation and research concerning the secondary effects of hibernation is still lacking. For example, the possible effects of hibernation on memory, the metabolism or the immune system are unknown. Another problem is the deleterious effects of zero gravity combined with the inactivity of hibernation, although this may be resolved by using some means of artificial gravity. Other challenges include the problems associated with how the hibernation state is induced, established, regulated and exited and how administration of compounds to a hibernating human can be achieved. Achieving and perfecting human hibernation will require expertise in and integration of pharmacology, genetic engineering, environmental control, medical monitoring, AI, radiation shielding, therapeutics, spacecraft engineering and life support. Once all these disciplines have been successfully integrated human hibernation will make long-haul spaceflight a little more comfortable.

Bioethical training

Currently, there is no ethical framework to guide mission planners, mission commanders or crewmembers in their decision-making when it comes to answering some of the potentially awkward moral questions posed by interplanetary exploration (Table 7.5).

Perhaps the most discussed ethical aspect is the issue of how to cope with sexual desire among a crew of healthy young men and women. Fortunately, although sex has long been almost a taboo topic at NASA, the question of how to address sexual desire is probably the easiest to solve by simply having an all male or all female crew. Although such a selection policy may deselect candidates who may be better qualified, the behavioral issue that sex poses on such a long mission outweighs any

PLATE 1 A whole disk image of the planet Mars taken by the Mars Global Surveyor spacecraft in June 2001. It is winter in the southern hemisphere and there is dust storm activity in the Tharsis volcanic region. (Malin Space Science Systems/NASA.)

PLATE 2 Orion is NASA's replacement for the Space Shuttle and is due to enter service in 2015. It will also be the vehicle carrying Mars crews to low Earth orbit. (NASA.)

PLATE 3 NASA's Ares V is a two-stage, vertically stacked launch vehicle capable of carrying 188 metric tonnes to low-Earth orbit. For the initial insertion into Earth orbit, the Ares V first stage utilizes two five and a half segment reusable solid rocket boosters derived from the Space Shuttle's solid rocket boosters. In the Mars Direct architecture, the Ares V would deploy a forty-tonne cargo payload into a direct trans-Mars trajectory. (NASA.)

PLATE 4 The Skylon Single-Stage-to-Orbit (SSTO) spaceplane in orbit. (Adrian Mann, Reaction Engines Limited.)

PLATE 5 Aerocapture is a flight maneuver used to insert a spacecraft into orbit using the atmosphere as a brake. The atmosphere creates friction, used to slow the vehicle, transferring the energy generated by the vehicle's high speed into heat. The maneuver enables quick orbital capture without the requirement for heavy loads of propellant. (NASA.)

PLATE 6 US astronaut Franklin Chang-Diaz working on the International Space Station during Space Shuttle mission STS-111. (NASA.)

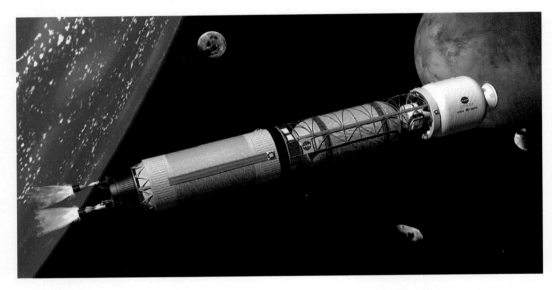

PLATE 7 Artist's rendering of Bimodal Nuclear Thermal Rocket. (John Frassanito and Associates/NASA.)

PLATE 8 An antimatter spaceship. (NASA.)

PLATE 9 J-2X rocket engine. (NASA.)

PLATE 10 Orion prototype/manufacturing demonstration unit. (NASA.)

PLATE 11 The Airbus A300 performing parabolic flight. (Novespace.)

PLATE 12 Ares I launch. (NASA.)

PLATE 13 Astronauts drilling for samples. (John Frassanito and Associates/NASA.)

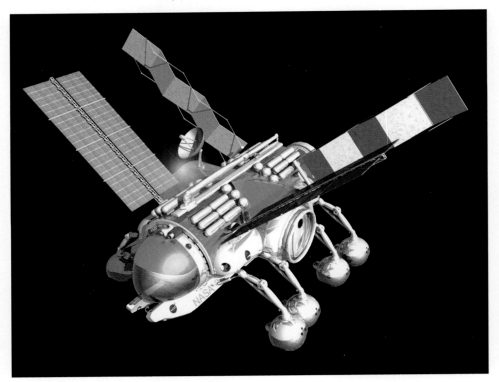

PLATE 14 EXP-Arch Mother Ship Rover/Scorpion. Side view. Design: Trotti Studio. (Mitchell Joachim.)

Table 7.5 Bioethical considerations for a manned mission to Mars

How do you get rid of the dead body of a crewmember?
When should life support be discontinued for a critically ill astronaut consuming valuable oxygen and endangering the remainder of the crew?
Should a straitjacket be included in the medical supplies?
Should NASA mandate prophylactic surgery such as removal of appendix, tonsils and gall bladder before a mission to Mars?
If a crewmember becomes disabled during the mission, who becomes their surrogate decision maker? Their spouse? NASA physicians? Other crewmembers?
Should NASA mandate genetic screening as a part of the astronaut recruitment process?

such disadvantages. As to the question of how to cope with sexual desire, the precedent set by innumerable multi-year polar expeditions provides overwhelming evidence that spending time in a confined environment with a group of men for several years under extreme duress does not result in any mission-compromising behavioral issues.

More pressing ethical questions concern the action to take in circumstances such as an astronaut becoming terminally ill during a mission. In such an event, the Commander may be directed by mission control to euthanize the ill crewmember in order to preserve medical supplies and life support consumables. Alternatively, the affected crewmember, knowing he/she has only a short time to live, may offer to sacrifice his or her life for the mission. In such a situation, what do the mission guidelines instruct the Commander to do? Needless to say, euthanizing a crewmember is a decision guaranteed not to endear any space agency to the media or the public who assume a person's well-being will always take precedence over mission success. Part of the reason for such a perception is that, to date, any astronaut becoming crucially ill or injured onboard the ISS can simply leave the orbiting outpost and return home within hours onboard the Soyuz. Unfortunately, this will not be possible when the nearest hospital is several million kilometers away. For situations such as a crewmember dying, and other unforeseen eventualities, it will be necessary to have guidelines directing the Commander.

Some of the solutions to these issues may sound severe and uncompromising but an expedition to Mars is one fraught with risk and therefore requires crewmembers to accept these risks in the same manner Shackleton and his crew accepted the risks of traveling to Antarctica.

ANALOG ENVIRONMENTS AS TRAINING TOOLS

Throughout the history of space exploration, analog environments have played a vital role not only in defining and designing space missions but also in training astronauts. Given the potential of high-fidelity simulations to prepare astronauts to conduct exploration and science activities, it is inevitable Mars-bound crews will spend a significant part of their training in one or more of the analog environments described here.

Antarctica
Apart from its rich history of exploration and the expeditions of Douglas Mawson and Ernest Shackleton, making it an appropriate location for an analog, Antarctica has a number of characteristics making it a serious candidate for use as a planetary simulation facility.

   Firstly, the continent of Antarctica is remote and hostile, as evidenced by the fact the United States Antarctic program has experienced more than sixty deaths since 1946. However, with proper logistical support and safety awareness, training can be conducted with acceptable levels of risk.

   Secondly, the isolation of the continent serves as an ideal analog for studying human factors and small human populations over long periods of time. During the winter months, those living on the continent are completely shut off from the rest of the world except for radio communications and the occasional airdrop resupply. What better place to simulate a Mars mission?

   Thirdly, the Antarctic Treaty provides a proven framework for international cooperative exploration and scientific efforts similar to the model envisaged for a human mission to Mars.

   Although not as isolated as the Antarctic, the polar region of Canada has played host to a planetary simulation facility for more than ten years. Designed to serve as an analog for Mars, the Haughton Mars Project (HMP), supported by both the American and Canadian space agencies, is perhaps the most successful and widely-known planetary analog.

Haughton Mars Project
The HMP is an international interdisciplinary field research project conducted in the vicinity of the Haughton impact crater on Canada's Devon Island. The project, conducted jointly by the Mars Institute, NASA, Search for Extraterrestrial Intelligence (SETI) and the Canadian Space Agency (CSA) utilizes the Mars-like features of the island and the crater to develop and evaluate new technologies and field operating procedures that may one day be used on a manned mission to Mars.

Crewmembers
Each summer, dozens of researchers, students, support staff and media journey to the HMP site. Some spend the entire summer there, while others rotate in and out for shorter stays of a few days or couple of weeks. In 2008, HMP celebrated its twelfth

field season featuring scientists and researchers from Simon Fraser University (SFU) in Vancouver, NASA's Ames Research Center (ARC), the CSA, the Mars Institute, Hamilton Sundstrand, McGill University and the National Space Biomedical Research Institute (NSBRI). Most crewmembers arrive courtesy of Marine C-130 crews, who support HMP with the delivery of expeditionary equipment, research gear and exploration vehicles. Getting around the HMP site is either by all-terrain vehicle (ATV) or simply by walking, often while wearing a spacesuit similar to the one that may be used on the Martian surface.

Environment

Although the Haughton crater is a cold, dry, barren, dusty and windy place with an average temperature of only –17°C, in comparison with Mars, which has an average temperature of –60°C and is drenched in lethal radiation, the HMP analog is a comparatively pleasant place. However, while elements of the Martian surface environment may be absent, the crater is a step in the right direction to evaluating strategies that will help plan the future exploration of Mars.

The Haughton Crater is characterized by ancient lakebeds, rocky terrain, occasional valleys and steep-walled canyons. Although rocks on Mars may have a different composition, the physical properties of the impact deposits still provide a valuable analog for the scientists stationed there. Recent neutron spectrometry data from the Mars Odyssey spacecraft has provided evidence of ice deposits at shallow depths in the Martian subsurface similar to subsurface ice at the Haughton Crater.

In addition to these subsurface ice deposits, Haughton Crater provides remnant signatures of hydrothermal activity and networks of channels bearing a close resemblance to the small valley networks on Mars. Perhaps the most valuable attribute of the Haughton Crater is the presence of life. Despite the high ultraviolet radiation environment during the summer and the harsh weather during the winter, Haughton Crater microorganisms are able to survive and adapt, a capability with potential implications for the search for life on Mars.

The HMP provides a valuable micro-scale experience of what it might be like to live on Mars. Not only has it successfully promoted the idea of traveling to Mars, the HMP has demonstrated such a forward-looking enterprise provides an ideal opportunity for international cooperation and also presents future Mars crews with a useful analog for the purposes of training. In fact, the HMP has been so successful it has led to the creation of similar analogs, such as the Mars Desert Research Station (MDRS).

Mars Desert Research Station

The MDRS is an analog to a Mars surface habitat located in the southwest desert region of the United States. Similar to the HMP, the MDRS was constructed for mission simulations according to Mars Design Reference Mission (DRM) guidelines [10]. One of the study approaches adopted during mission simulations is *exploratory methodology*, involving observing, recording, describing and analyzing an analog mission inside the habitat. Employing this methodology, investigators seek answers to questions such as how plans develop and change during a mission, how individuals interact during the day, and how public and private space is used. Other questions focus on human factors considerations such as the roles of the crew, requirements for life support and alternate operations concepts. Human factors questions are asked in an attempt to identify levels of concern relating to the habitat design. An example of a MDRS mission design is provided in Table 7.6, which summarizes the mission design options during the fifth simulation in April 2002.

Table 7.6 MDRS Mission Options [5]

Mission Option Criterion	MDRS Description	Alternative(s) considered for Mars mission
Crew location	Southwest Utah desert. Simulating habitat on Martian surface	In Mars orbit
Number of crew	Six	Four
Crew gender	Four men. Two women	All men or 50:50
Crew Structure	Commander (Cognitive Scientist) Health & Safety Officer (also Chief Engineer) Science Specialists (biologist, geologist, geophysicist)	Pilot Medical Doctor Aerospace Engineer
Accommodation	Staterooms with work areas	Minimal sleeping area
Duration	Two weeks	Six months
Life Support System	Open-loop	Recycling gray water Closed loop
Tasking, scheduling & control	Crew does all planning Mission support provides logistics assistance and schedules activities Sleep time up to individuals	Remote team dictates daily schedule with individual tasks and personal activities Sleep according to timeline
Communications	Daily commander, engineering and health/safety reports Detailed EVA reports Weekly science reports	Reports written by mission support team
Mission timeline	General planning in month preceding Crew did not meet prior to simulation	One year prior training and working together
Crew safety	Fire and medical emergencies Flight surgeon on call	Radiation and deconditioning dangers

During the fifth MDRS mission, one of the objectives was to characterize how space was used. Using a time lapse video technique, a complete day was analyzed to determine presence and absence of crew and the percentage of time spent in each area. This method helped researchers identify a work system within the habitat and provided a framework showing crew productivity and a means to study the use of time and space in the habitat.

NASA's Extreme Environments Mission Operations Project

"As an analogue to the International Space Station, the Aquarius habitat is just about perfect. When we are inside, it feels like we are onboard a remote outpost."

Astronaut, Mike Fincke, NEEMO Aquanaut.

Aquarius is a unique space analogue research station anchored sixteen kilometers from Key Largo in the Florida Keys National Marine Sanctuary. The facility, the only undersea laboratory in the world, provides living quarters permitting scientists and resident astronauts to stay on the seafloor for extended periods. The air inside Aquarius is pressurized to counterbalance the weight of surrounding water, which means crewmembers must breathe pressurized air. Breathing pressurized air for several days means the aquanauts are saturated with nitrogen, requiring them to perform a lengthy sixteen hour decompression to the surface at the end of their two week stay.

Undersea missions

Given an environment that so closely resembles the living and working conditions astronauts face in space, it is not surprising a NASA project, the NASA Extreme Environment Mission Operations Project (NEEMO), has used Aquarius on several occasions to conduct two-week missions and extravehicular activity (EVA) simulations. The two-week missions are planned in a very similar manner to a Space Shuttle flight, with each crewmember being assigned a detailed timeline of activities and specific timeslots for each activity, including outreach, sleep, chores and hygiene. One of the many activities performed by NEEMO's aquanauts are dives, planned analogous to space EVAs, requiring advance planning and a series of objectives, timelines and goals. While conducting underwater EVAs, astronauts practice communication tasks such as ship-to-ship calls to the International Space Station (ISS) and construction assignments such as building solar arrays and installing cables.

Exploration operations

During their two-week undersea increments, crewmembers have the opportunity to practice all sorts of operations and techniques that may one day be conducted for real as part of a manned Mars mission. For example, astronauts practice operating remotely operated vehicles (ROVs), providing training similar to that required for operating rovers on the Martian surface. They also go on the occasional treasure hunt, trying to figure out the best way to perform search and rescue (SAR). The treasure hunts begin with a crewmember dropping markers at random locations along the reef. Using Doppler navigation and transponders, crewmembers then go outside the habitat and try to find the markers using various search techniques.

Although only the size of a school bus, the cramped confines of Aquarius serve as a high-fidelity simulation of the cramped confines astronauts will occupy en-route to the Red Planet and will surely feature as a part of pre-mission training.

Mars500

The European Space Agency (ESA) has already gained experience in isolation and confinement studies. In 1990, the agency conducted the thirty-day Isolation Study for European Manned Space Infrastructures (ISEMSI), which was followed by the sixty-day Experimental Campaign for European Manned Space Infrastructures (EXEMSI) in 1992 and the Human Behavior in Extended Spaceflight (HUBES) study in 1994. Recently, the agency prepared for a study to simulate a five-hundred-day mission to Mars, dubbed Mars500.

The aim of the cooperative study between ESA and the Russian Institute for Biomedical Problems (IBMP) is to seal six carefully selected candidates inside an isolation chamber for one hundred and five days commencing in October 2008. This period will then be followed by the full isolation period of 520 days, commencing in early 2009. One section of the isolation chamber simulates the spacecraft that would transport them to and from Mars, and another section simulates the landing module that would transfer them to and from the surface of Mars.

> "We have seen many motivated, gifted and intelligent people over the past week and it will be a complex task in choosing the most suitable candidates for the study. Not only are we looking for robust, emotionally stable, motivated team workers who are open to other cultures and can deal with the slightly Spartan lifestyle you would associate with an actual space mission, we also need to combine different personalities and talents together in order to create the optimal group for such an extensive exercise."
>
> *Henning Soll, Psychologist, German Space Agency.*

The study attracted 5,600 applicants, of whom thirty-two were invited to the European Astronaut Centre (EAC) in Cologne, Germany, on May 27, 2008, for further evaluation, including a medical examination, a psychological evaluation and a personal interview with an expert panel to determine the suitability of each candidate.

The experimental facility, which includes the isolation facility, operations room and technical facilities, is located in a building of the IBMP in Moscow. The layout comprises four hermetically sealed interconnected habitat modules and one external module, which will be utilized as an analogue for a Martian surface excursion. The volume of all habitat modules (Table 7.7) is five hundred and fifty cubic meters.

Once those lucky enough to be selected are sealed inside the isolation chamber, they will be subject to the same restrictions as an astronaut embarking upon a space mission. For example, voice contact with family and friends will be conducted via a simulated mission control centre and, as with all astronauts, they will be subject to extensive testing and evaluation. Scientific investigations will include analysis of urine, blood, ECG, sleep quality, and the influence of exercise and food supplementation, in an attempt to aid the development of potential countermeasure tools and techniques that may be implemented during the real manned mission to Mars.

IN SUMMARY

During the era of manned spaceflight, much has been learned about the selection and training of astronauts. Nevertheless, the challenges of a Mars mission will demand a re-evaluation of the selection guidelines and a re-assessment of the training required to prepare astronauts to cope not only with the stress of isolation and confinement but also with the harsh physical environment which is the subject of the following chapter.

Table 7.7 Mars500 experimental facility

Module	Dimensions (m)	Designation	Description
1	3.2 × 11.9	Technical-medical	Houses two medical berths, a toilet and equipment for routine medical examinations. Also includes equipment for performing telemedicine and diagnostic investigations.
2	3.6 × 20	Living Quarters	The main living quarters comprise six individual compartments, including a kitchen-dining room, main control room and lavatory. The 2.8 × 3.2 m crew compartments each have a bed, desk, chair and shelves.
3	6.3 × 6.17	Mars Landing Module	This compartment will be used only during the thirty-day Mars orbiting phase. It comprises three bunk beds, two workstations, a lavatory, control and data collection system, communications system, ventilation system, waste treatment facility, and a fire suppression system.
4	3.9 × 24	Storage Module	Comprised of four compartments Compartment 1 — Fridge Compartment 2 — Storage of non-perishable food. Compartment 3 — Location of experimental greenhouse Compartment 4 — Bathroom, sauna and gym.

REFERENCES

1. Ayre, M.; Zancanaro, C., and Malatesta, M. Morpheus – Hypometabolic Stasis in Humans for Long Term Space Flight. Journal of the British Interplanetary Society. Vol. 57. 2004.
2. Boyer, B., and Barnes, B.M. Molecular and Metabolic Aspects of Mammalian Hibernation. Bioscience, 49, No. 9, 1999.
3. Cater, J.P., and Huffman, S.D. Use of Remote Access Virtual Environment Network (RAVEN) for Coordinated IVA-EVA Astronaut Training and Evaluation. Presence: Teleoperators and Virtual Environments. Vol. 4. No. 2 (Spring 1995). p. 103–109.
4. Chung J.; Harris M.; Brooks F.; Kelly M.T.; Hughes J.W.; Ouh-young M.; Cheung C.; Holloway R.L., and Pique M. Exploring virtual worlds with head-mounted displays, non-holographic 3-dimensional display technologies, Los Angeles, 15–20 January 1989.
5. Clancey, W.J. Participant Observation of a Mars Surface Habitat Mission Simulation.
6. Diener HC,; Wist ER,; Dichgans J,; Brandt T. The spatial frequency effects on perceived velocity. Vision Res. 1976; 16:169–76.
7. Genetic Information Nondiscrimination Act. H.R. 493, May 21st, 2008.
8. Graf, J.; Finger, B.; Daues, K. Life Support Systems for the Space Environment. Basic Tenets for Designers. http://advlifesupport.jsc.nasa.gov/. 2001.
9. Hayes-Roth, B., and Larsson, J.E. Guardian: an intelligent autonomous agent for medical monitoring and diagnosis. IEEE Intelligent Systems (January/February), pp. 58–64, 1998.
10. Hoffman, S.J., and Kaplan, D.I. (eds). Human Exploration of Mars: The Reference Mission of the NASA Mars Exploration Study Team. NASA Special Publication 6107. Lyndon B. Johnson Space Center, Houston, Texas. (Addendum, Reference Mission Version 3.0, June 1998, EX13-98-036). 1997.
11. Rebo R.K. and Amburn P. A helmet-mounted environment display system. In Helmet-Mounted Displays, SPIE Proceedings vol. 1116, pp. 80–4. 1989.
12. Slater, M.; Steed, A.; McCarthy, J.; Maringelli, F. The Influence of Body Movement on Subjective Presence in Virtual Environments. Human Factors, 40 (3), 469–477. 1998.
13. Stanney K.; Mourant R.; Kennedy R.S. Human Factors issues in Virtual Environments: a review of the literature. Presence 7: 327–351. 1998.
14. Stanney K. Handbook of Virtual Environments: Design, Implementation and Applications. New York: Lawrence Erlbaum Associates.

8

Biomedical and behavioral issues

The annals of polar exploration are replete with accounts of explorers suffering from scurvy, starvation and vitamin toxicity. A number of these expeditions were also afflicted by depression, lethargy and disabling psychological events. While the dangers of malnutrition were an ever-present and, occasionally, life-threatening risk for early explorers, medical dangers such as radiation, bone demineralization and psychological dysfunction present risks equally, if not more, hazardous to those embarking on a voyage to the Red Planet. While the aforementioned medical conditions suffered by early polar explorers were caused by a lack of knowledge concerning nutrition and group dynamics, a similar fate may befall Mars explorers because scientists aren't sure how crewmembers will be affected by radiation, bone loss or how they will respond to being isolated for more than two years. To better understand these risks, this chapter deals first with the primary biomedical risks and concludes with an assessment of the behavioral problems that may be encountered.

BIOMEDICAL RISKS

The first part of this chapter discusses the primary biomedical risks (Table 8.1) faced by those embarking upon a voyage to the Red Planet. While almost all the issues listed in Table 8.1 will manifest themselves over the course of such a mission, it is beyond the scope of this book to discuss each of them in detail. Therefore, the focus of this chapter is to describe the cumulative effects of the potentially most devastating issues, namely those associated with radiation and bone demineralization.

RADIATION

Overview
It is more than thirty-five years since astronauts ventured beyond Earth's protective magnetic shield and travelled through interplanetary space to the Moon. Although the Apollo missions subjected astronauts to space radiation, the short duration minimized the risk, but a Mars mission will subject crewmembers to prolonged

Table 8.1 Biomedical risk factors during a mission to Mars

Radiation Exposure	Bone Loss
Synergistic effects from exposure to radiation and microgravity.	Injury to soft connective tissue, joint cartilage and intervertebral disc.
Damage to central nervous system.	Renal stone formation.
Early or acute effects resulting in increased risk of cancers.	Fracture and impaired fracture healing.
Effects on fertility and heredity.	Development of osteoporosis.

Muscle Loss	Cardiovascular
Loss of muscle mass, strength and endurance.	Impaired cardiovascular response to orthostatic stress.
Proneness to muscle injury.	Diminished cardiac function.
Impact of degeneration of muscle or increased injury of muscle.	Impaired cardiovascular response to stress.
Loss of motor control.	Occurrence of cardiac dysrhythmias.
Neurovestibular	Immunological
Re-entry and landing vertigo.	Immunodeficiency/Infections.
Post-landing imbalance, vertigo and visual instability.	Altered hemodynamics caused by altered blood components.
Inflight spatial disorientation. Vestibular function changes.	Altered wound healing. Altered host-microbial interactions.

exposure to the deep space radiation environment. Such a long exposure to space radiation will impose risks upon the crewmembers such as an increased chance of carcinogenesis, acute and late central nervous system (CNS) disorders, and chronic and degenerative tissue risks [25, 29]. While mission planners will do their best to provide countermeasures and a storm shelter for those travelling to Mars, protecting crewmembers from the effects of extremely energetic galactic cosmic radiation (GCR) may prove impossible.

Radiation environment
Astronauts bound for Mars will be exposed to two types of ionizing radiation capable of inflicting serious and potentially lethal effects.

The first of these are *heavy ions* (atomic nuclei with electrons removed) associated with GCR, consisting of approximately 87% protons, 12% alpha particles and 1% high energy (HZE) particles. These nuclei have been accelerated to extremely high energies outside the Solar System and are capable of producing harmful effects in astronauts.

Solar particle events (SPEs) produce energetic protons with high energies capable of delivering high-dose rates to insufficiently shielded crewmembers.

The two largest SPEs observed were in August 1972 and October 1989 [26]. Such events are used to provide realistic estimates of the SPE environment that may be encountered during missions taking place during active solar conditions and may also assist in the prediction of SPEs. For example, the October 1989 SPE was predicted by the National Oceanic and Atmospheric Administration (NOAA) as a result of an X-ray burst occurring one hour prior to the SPE onset. Although the actual event was predicted successfully, the severity of the SPE was not forecast with much success.

Radiation in deep space

Although much has been written concerning the long-term effects of the cumulative radiation dose and the increased risk of cancers in astronauts, mission planners should perhaps be more concerned with dose levels that trigger short-term radiation sickness. A major unexpected solar flare event could potentially disable the entire crew and with the long time lags between the vehicle and mission control, and no abort capability, crewmembers could face a truly unpleasant end to their lives in the remote isolation of deep space.

The radiation environment encountered by astronauts en-route to Mars will be characterized by solar activity. Solar activity cycles last between ten and twelve years and with each activity cycle there is a period of active solar conditions lasting between three and four years. The greatest probability of a SPE occurs between the rise (to solar maximum) and decline (to solar minimum) of solar activity, whereas the magnitude of GCR flux is greatest during solar minimum. Nevertheless, during the transit to and from Mars, every cell of an astronaut's body will be hit by protons or secondary electrons every few days and by a high energy ion about once a month. Gradually, whole-body doses will accumulate, and the risk to health will increase. If astronauts are lucky, the accumulated dose may still be within career limits by the time they eventually return to Earth. However, if a solar flare event occurs, the possibility exists that the crew may not survive.

To provide crews with as much warning as possible, scientists will monitor the solar weather. One feature they will pay particular attention to will be sunspots, which form where intense magnetic fields twist and poke through the surface. Sometimes, when these magnetic field lines twist to the point of snapping, enormous amounts of stored energy are suddenly released into the Sun's outer atmosphere. The resulting eruption is called a solar flare, which accelerates subatomic particles to enormous speed and spews out ultraviolet (UV), X-ray and gamma-ray radiation into space. Often, solar flares are followed by a coronal mass ejection (CME), in which billions of tonnes of the Sun's plasma are ejected into space travelling at more than fifteen hundred kilometers a second (Figure 8.1). Depending on how far from the Sun the astronauts are located, the radiation pulse from the flare may take anything from eight to twenty or more minutes to arrive. With no atmosphere shielding the vehicle, the spacecraft will be bombarded by the radiation pulse, affecting communication capability and electronic equipment onboard.

Figure 8.1 In this 2002 SOHO image of a coronal mass ejection (CME) the Sun's face has been replaced by a simultaneous ultraviolet image. The solar environment out to a million kilometers beyond the Sun is shown, but the CME still extends beyond the limit of the frame. Near solar minimum CMEs occur about once per week, but they can occur twice a day at solar maximum. (SOHO Consortium, ESA, NASA.)

Martian atmosphere

Once the astronauts arrive at Mars, radiation hazards will be reduced thanks to the increased distance from the Sun and the protective effect of the atmosphere, although crewmembers will still require a storm shelter in the event of solar flare.

The amount of protection provided by the Martian atmosphere, with an average surface pressure of 0.6 kPa, changes seasonally as the surface pressure on Mars varies. Protection also fluctuates according to atmospheric pressure, which varies according to altitude, from just 0.03 kPa on *Olympus Mons* (24,000m) to 1.155 kPa in the six-kilometer depth of *Hellas Planitia*.

The Committee on Space Research (COSPAR) has used atmospheric data gathered during the Viking missions to develop an atmosphere model of Mars which may help mission planners predict radiation doses. The models, which used average daily temperatures and pressures at mid-latitude sites during the summer season, provide mission planners with an estimation of the amount of protection provided by the Martian atmosphere during different environmental conditions. Using this information, planners can calculate dose estimates on the Martian surface during worst-case scenarios such as the 1989 SPEs event. Based on the COSPAR models, the radiation doses do not approach astronaut career limits, even when large flare doses are added to the average GCR doses. However, while a surface stay on Mars may not appear to be damaging to health, the doses incurred during the transit to and from the Red Planet are likely to dominate the total mission dose. To understand how much radiation astronauts may be exposed to it is first necessary to be familiar with how radiation is measured.

Radiation units
The impact of high-energy radiation on humans is measured by the dose equivalent, which describes the effect of radiation on tissue and is calculated by multiplying the *absorbed dose* by the *quality factor*.

The absorbed dose, which describes the deposited energy of radiation in the body, is measured in *grays* (1 joule of energy absorbed in one kilogram of matter). On Earth, a human experiences less than 5×10^{-3} gray (Gy) in a year whereas a cancer patient receives approximately sixty Gy during a full course of therapy.

The quality factor is a measure that takes into account the relative effectiveness of the radiation in producing the biological effect. For example, it is known ionizing radiations such as protons, beta particles and energetic ions of heavier elements cause more biological damage than radiations such as X-rays and gamma-rays. The more damaging radiations are said to have a *Relative Biological Effectiveness* (RBE) greater than 1.0.

The RBE is defined as the ratio of the dose of a particular radiation to that of the test radiation required to cause an equal biological effect. For example, if 1 Gy of beta particles kills the same number of blood cells as 2 Gy of X-rays, the RBE of the beta particles is 2.0. If the 1.0 Gy of beta particles is multiplied by the RBE factor of 2.0, the biological equivalent dose is 2.0 Sieverts. Sieverts are used because they are the SI unit of biological equivalent radiation dose. It sounds a little convoluted but it is important to understand the difference between the two since the RBE concept is used in defining space radiation health.

Biological risks
For the purposes of determining human protection standards, the biological effects of radiation have been divided into *stochastic* and *deterministic* effects.

Stochastic effects

Stochastic effects are due to radiation-induced changes randomly distributed in the DNA of single cells that may result in cancer, depending on the cells affected. For stochastic effects, there is no threshold dose below which it is relatively certain an adverse effect cannot occur, but with prolonged exposure to even low dose radiation, the probability that cancer or some genetic effect will occur increases. However, cancers that may be caused by stochastic effects usually occur following a long latent period, often two or more years after the exposure to radiation.

Deterministic effects

While stochastic effects are the most important consideration in setting protection limits for humans exposed to radiation at low doses, *deterministic* effects (formerly known as nonstochastic effects) occur only after exposure to relatively high doses. These effects may range from acute radiation sickness (ARS) to nausea and, in contrast to stochastic effects, may occur within hours of exposure.

Early systemic effects

For fifty percent of individuals, prodromal radiation effects occur within a few hours of an acute whole body exposure in the range of 1.5 to 2.0 Gy and are characterized primarily by nausea and vomiting, although some individuals may experience nausea at doses in the range of 0.5 to 1.0 Gy. For doses between 2.5 and 3.0 Gy, almost all individuals experience nausea and vomiting.

Total Body Irradiation-Acute Radiation Syndrome (IARS) is characterized by a depletion of white blood cells (WBCs) and platelets. The threshold for these effects, which usually occur two to four weeks after exposure, is in the range of 1.5 and 2.0 Gy. Of particular relevance to Mars astronauts is the risk of being exposed to doses in the range of 3.0 to 4.0 Gy since, at this level of exposure, with minimal supportive care, death may occur.

Organ function

If too many cells of a certain tissue die, organ function will be compromised. For example, if cells lining the gastrointestinal tract die in sufficiently large numbers, the gut will be unable to absorb food or maintain electrolyte balance. This is the reason why, after suffering a large radiation dose, victims experience nausea and vomiting. However, cells do not have to die for organ function to be disrupted. Ionizing radiation may injure cells via many different pathways, depending on the sensitivity of a given tissue. For example, if full repair of cells fails, but not to the point of leading to the death of subsequent generations of cells, the damaged cells may survive and transform into cells that can become cancer precursors. Alternatively, damaged cells may lose some functional characteristics, in turn leading to organ failure.

Risk to fertility

Prolonged exposure to deep space radiation may also result in reduced fertility or transient or temporary sterility, lasting from several months to several years. In some cases, if the exposure is particularly prolonged or severe, the sterility may be permanent.

For males, the radiation dose required to cause temporary sterility is between 0.5 and 4.0 Gy [17], although a single acute dose of 0.15 Gy has been reported to cause a decrease in sperm count [10]. Temporary sterility may last from several months to several years [17]. Doses in the range of 2.5 and 4.0 Gy may cause permanent sterility, but infertility may also be caused by low-dose-rate protracted exposure [16]. Given the prolonged exposure to deep space, and given there is a significant likelihood of Mars astronauts approaching or exceeding career radiation limits, it is the latter process that is probably the most likely to sterilize crewmembers.

A radiation dose in the range of 6.0 to 20.0 Gy is required to sterilize females, whereas temporary sterility may occur at doses between 1.25 and 3.0 Gy [5]. Doses of 2.0 to 6.5 Gy are required to sterilize five percent of females for more than five years.

Unless there are significant developments in radiation shielding between now and a manned mission to Mars, the risk of reduced fertility represents such a risk NASA may impose age restrictions on crewmembers or require astronauts of reproductive age to bank sperm or eggs. In either case it will be interesting to see how this problem is dealt with by space bioethicists.

Late effects
The primary late effects of exposure to radiation during an interplanetary mission include induction of cancer, CNS damage and genetic mutations.

Radiation effects on DNA
The potential risks to crewmembers' offspring as a result of exposure to deep space radiation are unknown. Consequently, it is probable Mars crews will receive counseling regarding the possible genetic effects of a three-year exposure to interplanetary radiation.

When the human body is exposed to radiation, the energy from the radiation is deposited at the cellular level by interactions between the radiation and the electrons of molecules comprising the cells. The deposition of radiation results in the atoms that make up complex molecules losing electron bonds that tie them to the molecule. In certain cases, the molecule will recover, but if the radiation deposition continues unabated, template molecules such as DNA, may be unable to repair the damage and may die. Alternatively, cellular repair mechanisms may be unsuccessful and leave damaged DNA cells incompletely repaired or lead to dying or aberrant cells in subsequent division. Such an unstable cell and its progeny will result in a little understood process known as *genomic instability*. Genomic instability is a hallmark of cancer cells and is thought to be involved in the process of carcinogenesis. In the hostile environment of deep space, radiation may induce such a process of instability, resulting in the multiple gene mutations necessary for the development of cancer.

Recent research [9] exposed DNA in solution to high energy radiation to investigate the types of damage produced as a result of the exposure. The results were not encouraging. The exposure resulted in multiple damage types, the most dangerous being damage clusters, which may cause genetic mutations and cancers.

In terms of heritable effects, the end points of exposure to radiation may include major congenital malformation, stillbirth and tumors with onset prior to the age of twenty.

Fortunately, some individuals possess genotypes that confer upon them an increased resistance to radiation. A genotype is simply a set of physical DNA molecules inherited from parents and to reduce the chances of genetic instability among its crewmembers, it is possible NASA will select only those astronauts with reduced susceptibility genotype. Again, this becomes a question of bioethics.

While the genetic outlook for Mars-bound astronauts may appear less than rosy, the repair mechanisms utilized by the human body to repair DNA after exposure to radiation are extremely versatile, and sophisticated cellular processes exist for repairing all types of DNA damage. These processes are capable of repairing not only base damage but also single and double-strand breaks. Nevertheless, given the risk of adverse biological effects that may be inflicted by ionizing radiation, this is one particular area that demands more research.

Cancer risks

The most likely deleterious biological result of irradiation during a manned mission to Mars is late-occurring cancers. Cancer risks are calculated using the absorbed dose from radiation, according to information provided by the National Council on Radiation Protection and Measurements (NCRP). Unfortunately, these calculations are based on a function of a multitude of factors subject to varying degrees of uncertainty. Given the vagueness of the data, defining a calculated risk of cancer as a result of absorbed radiation is a challenge and the risk can only be estimated using a probability distribution function (PDF).

Other sources of data mission planners use to help reduce uncertainty include animal systems and cellular studies, although data on tumor induction following exposure to protons and heavy ions remains sparse. The few cellular studies conducted used protons of different energies to examine cell survival and induction of chromosomal aberrations [19, 27]. Although the range of energies used in these studies was lower those encountered during an interplanetary expedition, the data suggested similarities in effects between protons and gamma rays. More importantly for the astronauts being irradiated, the studies indicated evidence for repair of proton-induced DNA damage.

Central nervous system effects

Based on research to date, it is suggested much of the damage inflicted by protons upon the CNS will be repairable by DNA and regrowth of cells. Studies also indicate the cell nuclei of the CNS are radiation resistant to protons, although it is still possible protons may impair the functional capability of the CNS during the voyage to Mars.

More troubling are the effects of heavy ions, which may induce significant damage to the CNS due to the ability of these particles to traverse several layers of cells. The dense ionization of heavy ions inflicts not only cellular damage and biochemical changes, but also functional effects and disrupts the DNA repair process. Unfortunately, biological research investigating heavy ions is restricted to a few centers in the world with suitable accelerators, and the few experiments that have been conducted have proved challenging.

Cataracts

Cataractogenesis is the term given to the process of cataract formation. Cataracts are one of the myriad occupational hazards faced not only by Mars astronauts but those who work on the ISS and the Space Shuttle. While threshold doses exist for workers in LEO, there is no data for induction of cataracts in humans exposed to heavy ions, and only limited data for induction by protons [33].

Risks to brain stem cells

"The exceptional sensitivity of these neural stem cells suggests that we are going to have to rethink our understanding of stem cell susceptibility to radiation, including cosmic radiation encountered during space travel, as well as radiation does that accompany different medical procedures."

Dennis A. Steindler, Ph.D., Executive Director, McKnight Brain Institute,
University of Florida, December, 2007.

Unfortunately, the risk of cancer and genetic damage are not the only radiation risks faced by astronauts bound for Mars. Recent research conducted by scientists at the NASA Space Radiation Laboratory (SRL) and at the Brookhaven National Laboratory (BNL) in Upton, New York, indicates space radiation selectively kills stem cells in the area of the brain believed to be important for learning and mood control.

In the study, mice were administered a single dose of radiation equal to the amount astronauts might receive during a three-year return trip to Mars. Unexpectedly, the scientists found a special type of stem cell in the hippocampus is selectively killed by the radiation. Stem cells are important because they have the ability to renew themselves and produce many different types of cells.

"Space radiation has not been a serious problem for NASA human missions because they have been short in duration or have occurred in low earth orbit, within the protective magnetic field of the Earth. However, if we plan to leave low Earth orbit to go on to Mars, we need to better investigate this issue and assess the risk to the astronauts in order to know whether we need to develop countermeasures such as medications or improved shielding. We currently know very little about the effects of space radiation, especially heavy element cosmic radiation, which is expected on future space missions and was the type of radiation used in this study."

Philip Scarpa, M.D., NASA Flight Surgeon

Given the exceptional sensitivity of these neural stem cells to radiation and the fact the stem cells affected are in those areas of the brain associated with cognitive and emotional function, the results of the study present mission planners with a new problem.

Radiation Exposure Guidelines

NCRP

According to NCRP guidelines and the International Commission on Radiological Protection (ICRP), the maximum *annual* dose equivalent for the general public is one thousandth of a Sievert (mSv). In contrast, the maximum annual dose equivalent for nuclear workers is 50 mSv. The unit, milliSieverts, is commonly used to measure the effective dose in diagnostic medical procedures such as X-rays, nuclear medicine and computed tomography. For acute full body equivalent dose, one Sievert is sufficient to cause nausea, while three Sieverts will cause death in 50% of cases within thirty days.

> "Unlike LEO exposures, which are often dominated by solar protons and trapped radiation, interplanetary exposures may be dominated by GCRs, for which there is insufficient data on biological effects. Consequently, risk prediction for interplanetary space is subject to very large uncertainties, which impact all aspects of mission design."
>
> *ESAS Report*

NASA's Guidelines

To date, there are no guidelines for allowable radiation exposure in deep space. In a federally mandated obligation to follow the ALARA (As Low As Reasonably Achievable) principle of keeping radiation exposure as low as reasonably achievable, the guidelines NASA uses for its astronauts working on the International Space Station (ISS) are based on a point estimate[1] for the levels of radiation causing an excess risk of 3% for fatal cancer due to exposure (Table 8.2).

One way analysts [1] generate radiation doses for a manned mission to Mars, is to use the guidelines summarized in Table 8.2 and generate point estimates. One reason for this is because, computationally, the calculation of conventional exposures based on linear energy transfer (LET)[1] in a target medium such as the human body, may be performed with little ambiguity. Unfortunately, this method, which is basically an advanced form of guessing, is fraught with uncertainty and any allowable doses calculated by this method cannot be treated as a rigid requirement because the cell damage caused by ionizing radiation is highly variable for different cell types.

Calculating the radiation limits for interplanetary space

Rather than extrapolating the LEO to calculate deep space mission exposures it may be possible to use the computation procedure devised by Rapp [20], outlined here.

(i) Divide the mission into phases and assign each phase a duration. For example, Phase One would be the Transit Phase from Earth to Mars and would last 180 days.

---

1 Point estimation uses sample data to calculate a single value, which serves as a 'best guess' for an unknown population parameter.

Table 8.2 Radiation levels causing excess cancer risk

(A) Recommended organ dose equivalent limits for all ages [NCRP]

Exposure Interval	BFO[1] Dose Equivalent (cSv)[2]	Ocular Lens Dose Equivalent (cSv)	Skin Dose Equivalent (cSv)
30-Day	25	100	150
Annual	50	200	300
Career	See Panel B	400	600

1. Blood-forming organs.
2. centi-Sieverts

(B) LEO Career whole-body effective dose limits (Sv)[3] [NCRP]

Age	25	35	45	55
Male	0.7	1.0	1.5	2.9
Female	0.4	0.6	0.9	1.6

3. Sieverts

(ii) For each phase of the mission, define the energetic particle fluxes resulting from GCR and SPE.
(iii) Calculate the effect of the Martian atmosphere upon energetic particle fluxes to define the radiation arriving at the surface.
(iv) Estimate the energetic particle flux inside the Martian habitat.
(v) Convert the net flux into absorbed, equivalent and effective doses and compare the estimated doses with permissible doses for astronauts.

Likely dose for Mars astronauts
Based on these and other computational procedures, it is estimated during the 1000-day mission duration of a manned Mars mission the effective doses from GCR will be in the order of 1000 mSv. In addition, there is a significant probability for one or more SPEs. Furthermore, Mars crews will receive exposures from diagnostic X-rays, experimental protocols, air training and previous space missions. Astronauts selected for Mars missions will have some non-negligible radiation exposure history that will need to be considered into the risk management matrix for exploration mission crews.

Radiation countermeasures
Potential countermeasures are classified into the three categories of *operations*, *shielding* and *biological*. In this section, each countermeasure is discussed and an assessment of the dilemmas facing mission planners during their implementation is considered.

2 Linear Energy Transfer is a measure of the energy transferred to material as an ionizing particle travels through it. The measure is used to quantify the effects of ionizing radiation on the body.

"Protection against the hazards from exposure to ionizing radiation remains an unresolved issue. The major uncertainty is the lack of data on biological response to galactic cosmic ray exposures but even a full understanding of the physical interaction of GCR with shielding and body tissues is not yet available." [32]
NASA Technical Memorandum 4422 [32]

Operations

A unique aspect of a human mission to Mars is that there is no simple or fast abort-to-Earth option in the event of a radiation event. One way to reduce the risk of exposure to radiation events is to choose an optimal trajectory. As discussed in Chapter 3, the two principal classes of Mars mission trajectories are conjunction and opposition class.

A conjunction class mission is considered a minimum delta velocity trajectory requiring a transit time of between 350 to 550 days on the surface of Mars and one-way trip times between 200 and 300 days for a total round-trip time of between 900 and 1,000 days. Opposition class missions are characterized by short-duration stays of between 20 and 60 days, with one-way trip times of between 450 and 500 days. This latter class of trajectory is also characterized by asymmetric times between the inbound and outbound phases of the journey, with one phase being significantly longer than the other. Another aspect of this class of trajectory is a Venus swing-by, which would subject astronauts to more intense solar particle exposure. From the comparison of the considered mission trajectories, it is clear that, from a radiation exposure perspective, the long-stay fast transit trajectory is the safest option for a crewed Mars mission.

Shielding

The two options that may be considered for protecting astronauts from radiation in space include *active* and *passive* shielding.

Active Shielding

Active radiation shielding includes concepts such as electrostatic fields, plasma shields, and confined and unconfined magnetic fields. Studies investigating the efficacy of electrostatic shields revealed this method to be unsuitable for GCR shielding since the electrical considerations limited the minimum physical size of the shield to dimensions on the order of a few hundred meters [28].

Plasma shielding utilizes a large electrostatic field to repel positively charged particles but since this attracts and accelerates electrons to very high energies, a lower intensity magnetic field is used to control a cloud of free electrons, which deflect the incoming electrons. Although the concept appears attractive from a weight-saving perspective, the technology requires a system capable of achieving electrostatic potentials exceeding 200,000 kV on the surface of Orion [15].

A confined magnetic field utilizes concentric spheres arranged in a configuration designed to limit the spatial extent of the magnetic field to a limited region around the spacecraft. Although such a system has demonstrated efficacy as a shield against SPEs, the concept was not effective at stopping GCR.

Unconfined magnetic field shield concepts utilize a toroidal-shaped spacecraft. In this system, the spacecraft has a magnetic field generated by passing a current through coils. Unfortunately, such a system suffers from the same problem as the confined magnetic field and is unable to adequately protect against GCR.

Passive Shielding

The passive approach of bulk shielding is the approach NASA has adopted for the Constellation program.

The effectiveness of a shield material is determined by the transport of energetic particles within the shield. This process is characterized by the interactions of the local environmental particles with the nuclei and atoms of the shield material. One problem facing mission planners is choosing a shield material that not only protects astronauts against radiation but also possesses the qualities required to build space structures. For example, materials with high hydrogen content generally possess high shielding properties but do not have the qualities required for building a spacecraft due to the lack of structural integrity of the material. Presently, the material candidates of choice amongst mission planners are organic polymers. Other multifunctional candidate materials being considered include liquid hydrogen, methane and polyethylene (Table 8.3). These materials have been selected as candidate materials not only because of their shielding properties and structural integrity but also because of their density, another important consideration when it comes to launching missions elements into orbit.

Table 8.3 Chemical composition of candidate shielding materials

Material	ID	Density g/cm^2	Material	ID	Density g/cm^2
Aluminium 2219	ALM	2.83	Lithium Hydride	LIH	0.82
Polyetherimide	PEI	1.27	Liquid Methane	LME	0.466
Polysulfone	PSF	1.24	Graphite Nanofibres	GNF	2.25
Polyethylene	PET	0.92	Liquid Hydrogen	LH2	0.07

En-route to Mars, the Orion capsule and the TransHabs will be the principal shielded volumes for the crew. The capsule provides adequate shielding due to its structure, avionics, life support, consumables, waste storage and other hardware that protect crews from low-energy SPEs. The TransHabs water bladder also provides protection but for the rarer, high energy events, more protection is required. Since the duration of the most hazardous phase of an SPE or a close series of high-energy SPEs may last for hours or days, the Orion capsule or TransHab must be able to provide a storm-shelter capability for an extended period, during which the crew must have access to food, water and hygiene facilities.

Lockheed Martin considered a number of shielding solutions such as hull shielding, deployable water shielding and deployable high-density polyethylene (HDPE), with a density greater than 94 g/cm^2. At the time of writing, it is likely

Orion will have a deployable 2.5 cm thick HDPE shield stowed on the floor of the Environmental Control and Life Support System (ECLSS) and, in the event of a high-energy SPE, the crew will either configure a shelter inside the capsule or retreat inside the TransHab, depending on the protection afforded by the bladder.

Radioprotective agents

Vomiting and nausea are expected symptoms following moderate and acute exposure to radiation. In radiation medicine, antiemetic agents are routinely used to mitigate the effects of chemotherapy, but such an approach is unlikely to be adequate for crewmembers facing SPEs and GCR exposure.

One strategy being investigated to protect astronauts from the deleterious effects of radiation is the use of radioprotective agents, also known as *radioprotectors*, agents that protect cells from the damaging effects of exposure to ionizing radiation. Radioprotectors may be administered before and/or after radiation exposure and have been shown to operate by a variety of mechanisms, such as their antioxidant properties.

Amifostine and Melatonin

Recent research has produced new data contributing to the growing body of evidence confirming the efficacy of amifostine and melatonin as radioprotectors [12], although, at the time of writing, the drugs are only approved for clinical use. Amifostine, developed by the Walter Reed Institute, and also known as WR2721, is the only radioprotectant approved for use in humans and has a dose-reduction factor[1] of 1.8 to 2.7 when the drug is administered prior to an exposure. The drug has been on NASA's list of radioprotective drugs for several decades by virtue of its ability to inhibit the induction of mutations following radiation exposure and, thereby, reducing the risk of carcinogenesis.

Genistein

Genistein is an isoflavone with the characteristics of an ideal radioprotectant since it is non-toxic and is a natural product available in the diet from a single food source and may be administered daily to provide a long window of protective efficacy. Studies have demonstrated genistein's efficacy in protecting against radiation-induced lethality and enhancing survival when administered one day before radiation exposure.

Radiation forecasting

NASA's Space Radiation Research Program

One of the goals of NASA's Space Radiation Research Program (SRRP) is to provide astronauts with a means of predicting SPEs and to develop warning

---

1 Ratio of dose of radiation to cause an effect in the presence of the drug to the dose of radiation to cause the same effect in the absence of the drug.

strategies in the event of radiation events. Although solar radiation storms are notoriously difficult to predict, NASA's Solar and Heliospheric Observatory (SOHO) enables scientists to provide astronauts with a one hour warning of such an event, sufficient time to allow the crew to place the vehicle in safe mode and to retreat to their storm shelter. Although the warning system provided by SOHO is not perfect, the satellite, launched in 1995, has now operated through an entire solar cycle including the most recent solar maximum in 2001 and the dataset from these observations allow scientists to accurately forecast solar events. For example, scientists predicted all four major ion storms in 2003 with advance warnings ranging from seven to seventy-four minutes, a success rate that bodes well for future long duration missions.

BONE LOSS

"Bone demineralization during the Mars stay is unknown, but should be less than for an equivalent 0 g exposure. During the transfers the level of demineralization could reach 50% at the pelvis and it will certainly be more than the 15% threshold (considered as the level of significant increase of bone fracture risk). Bone demineralization is therefore an unacceptable risk, and must be controlled."
Statement from HUMEX-TN-001 paper (Definition of Reference Scenarios for a European Participation in Human Exploration and Estimation of the Life Sciences and Life Support Requirements, Sep 2000).

Bone demineralization is characterized by the rapid and continuous loss of bone mineral, alterations in bone architecture and alterations in skeletal mass which result in a condition similar to osteoporosis [31]. This microgravity-induced loss of bone mineral density (BMD) has been documented primarily in the weight-bearing components of the skeletal system such as the femur, pelvis and spine. Several research studies have corroborated data indicating astronauts may lose between 1 and 2% of their BMD per month [14, 18], a rate almost five times the rate of women with postmenopausal osteoporosis! To illustrate this, bone loss in cosmonauts during missions of up to 180 days is presented in Figure 8.2.

After just one year in space, astronaut's en-route to Mars may lose 20% of their BMD, equating to a 40% loss in bone strength [30]. The reduced gravity of Mars will lessen the impact upon the bones, but given the magnitude of bone loss, astronauts will be highly susceptible to risk of fracture. Furthermore, in the event of a crewmember suffering a fracture, healing would be inhibited due to the reduced gravitational field of Mars. Perhaps more worrying is the possibility astronauts returning from Mars may not only never fully recover their bone mass, but may suffer related health hazards such as toxic accumulations of excess mineral in the kidneys.

To better understand the hazards presented by bone demineralization, this section examines the physiological processes occurring in the skeletal system during

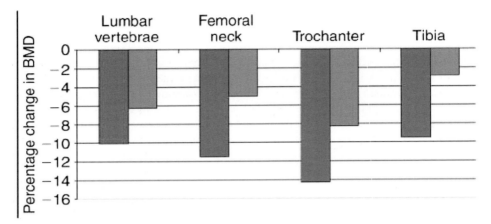

Figure 8.2 Percentage change in bone mineral density of femur for cosmonauts following missions of up to 180 days. For each of the four examples, dark grey and light grey represent maximum and average gain/loss in bone mineral density respectively. (AMPDXA for Precision Bone Loss Measurements on Earth and in Space. Harry K. Charles Jr., Michelle H. Chen, Thomas S. Spisz, Thomas J. Beck, Howard S. Feldmesser, Thomas C. Magee, and Barry P. Huang/ John Hopkins APL Technical Digest, Volume 25, Number 3 (2004), Page 188.)

extended spaceflight and the research directed at developing appropriate and effective countermeasures.

Effect of microgravity on the skeletal system
Exposure to microgravity produces adaptations in nearly every physiological system. Some adaptations, such as motion sickness, are self-limiting with symptom-resolution occurring within days, while other adaptations produce more progressive changes, the most serious being those imposed upon the skeletal system, which shows no signs of resolution.

One of the most regularly documented physiological changes associated with the spaceflight environment is the process of bone demineralization, caused by the absence of weight-bearing while in microgravity [21]. An absence of load removes not only the direct compressive forces on the long bones and spine, but also the indirect loading on these bones from the pull of muscles on the various bone structures to which they are attached. Invariably, the unloading of the skeleton leads to osteoporosis, weakening of the bones and delayed healing of fractures.

While the clinical implications of many of the physiological systems affected by microgravity have been extensively addressed, many are not yet fully understood. This is particularly true for exposure of the skeletal system to zero gravity for extended duration. For example, it is still a matter of discussion whether some of the observed demineralization changes may lead to permanent and irreversible damage. Of particular concern for Mars-bound astronauts is the uncertainty of knowing if bone loss will be replaced if exposure is extended beyond a tolerable threshold.

Mechanism of bone loss

Bone composition

Bone is composed of mineral and organic components. Collagen, the most abundant protein in bone, is synthesized primarily by osteoblasts and forms a framework upon which mineralization is superimposed. Adding to this process are various matrix proteins that have cell recruitment functions in remodeling bone.

At present there is little information concerning the influence of zero gravity on the biophysical functions of these matrix proteins. Compounding the issue is the suggestion of an impaired mineralization process that may occur during spaceflight. Bone demineralization is a complex and dynamic sequence of events involving mineral deposition regulated by cells responsible for aligning calcium phosphate crystals and depositing them within the collagen structure. Evidence from spaceflight indicates these minerals, when formed in microgravity, have a decreased crystal size and are configured imperfectly.

Bone cells

Also important to the understanding and prevention of bone loss during zero gravity are the processes of repair and remodeling. The remodeling process is governed by the osteoblasts and osteoclasts, although the control mechanism has not been identified. It is known that in space the astronaut's skeleton will experience a fast rate of resorption due to the unloading of mechanical stresses and weights. As a result of the skeleton no longer having to bear the astronaut's full weight, the body signals the osteoclasts to resorb bone at a fast rate and thereby begins to rid itself of what it believes is unnecessary bone. This process occurs in tandem with the rate of bone formation, which is negatively affected by zero gravity, resulting in a slowing of or destroying the action of osteoclasts and reducing the amount of calcium absorption. For example, on Earth bone absorbs 40 to 50% of the calcium intake, whereas only 20 to 25% is absorbed in space.

Bone homeostasis

Bone tissue is constantly recycled and renewed to maintain homeostasis, a process of bone-remodeling and repair resulting in approximately 500mg of calcium entering or leaving the bone each day. This remodeling occurs selectively in a process of reabsorbing or depositing bone tissue determined by the mechanical or gravitational stresses acting on the bone. The bone cells responsible for the removal of bone tissue are the osteoblasts while the ones responsible for bone repair are the osteoclasts. Together, these cells remodel bone tissue continuously in a process controlled by hormonal and mechanical feedback. Unfortunately, removal of gravitational stress results in a disruption to both of these processes, resulting in bone atrophying, a situation compounded by the effect of the blood supply.

Circulatory factors

All the physiological processes in bone depend on an optimal bone blood supply, but to understand how blood supply can cause bone atrophy we first need to be familiar with the effect of the absence of gravity on the circulatory system.

Gravity affects an organism *hydrostatically*, so when an astronaut is in an upright position the proportion of fluid volume in their lower half is greater than in their upper half. Once the force of gravity is removed however, the hydrostatic forces exerted on bodily fluid are completely neutralized and blood is distributed evenly throughout the body. This means the body detects less blood in the extremities such as the legs and the body's response to this unnatural blood redistribution is to pump more blood through the heart. Unfortunately, the increase in blood circulation leads to *accelerated* demineralization because increased blood flow results in an increased blood velocity through bone which increases the rate of calcium absorption into the blood supply.

Space radiation-induced bone loss
As if the physiological processes weren't bad enough, bone demineralization is exacerbated by the effect of radiation which results in a condition known as osteoradionecrosis.

Osteoradionecrosis, a condition of nonliving bone in a site of radiation injury) [2, 6, 8], and the denaturing of collagen fibers, has been observed in cancer patients receiving high doses of radiation during chemotherapy. Although research has not investigated the effect of ionizing radiation on general bone quality there is a high risk that Mars-bound astronauts may be exposed to sufficient radiation to cause significant decreases in both bone volume and bone integrity. These crewmembers will be exposed to high-energy heavy ions and unexpected SPEs and CMEs. While the fluence of heavy ions will be much less than that of protons, the energy deposition from individual particles will be much greater [15]. Also, since neither conventional shielding nor the Martian atmosphere will effectively protect the crew from cosmic radiation, astronauts will be exposed to a combination of secondary particle radiation and highly penetrating neutrons [2]. Although these astronauts will be exposed to lower levels than a cancer patient undergoing chemotherapy, when the planned two to three year duration of the mission is considered, it is possible crewmembers will be exposed to chronic and potentially health-threatening levels of radiation.

To assess the effect of radiation upon bone architecture during long duration missions, a recent study used microcomputed tomography to measure the effects of whole-body exposure to space-relevant radiation in mice [8]. Conducted at Clemson University, South Carolina, and the Brookhaven National laboratory in NASA's SRL, the study subjected groups of mice to radiation similar in intensity to that which Mars-bound astronauts will experience. Four months after exposure, the left tibiae and femurs were removed and analyzed by microcomputed tomography to determine architectural parameters such as bone volume, connectivity density, and trabecular spacing.

The results of this study were quite alarming in terms of the changes in bone architecture observed, since some of the changes suggested permanent deficits in bone integrity and the ability of the bone to sustain loading. It was suggested that although bone which had been exposed to space relevant radiation might recover bone mass, the ability and the efficiency of the bone to transmit loads may be permanently compromised. This is, in part, due to radiation killing stromal stem cells implicated in bone formation and growth [2]. Although Mars astronauts will be exposed to a lower dose rate of radiation than that investigated in the Clemson study, the cumulative dose will be a more important factor. Another factor that must be considered is the energy of the radiation and its effect on bone. In the study performed at Clemson University, the mice were subjected to monoenergetic radiation velocities, whereas during a mission to Mars, the radiation energies will be travelling at different velocities, a factor complicating the estimation of the dose astronauts may be exposed to.

Bone loss

Unsurprisingly, given the aforementioned problems, the study of bone deminer-alization has a long history. Studies conducted onboard the Russian space station, Mir (Table 8.4), revealed crewmembers lost more than one percent of bone mineral density per month, a figure similar to that experienced by astronauts onboard the international space station (ISS).

Given the rate of loss observed onboard Mir and the ISS, astronauts may risk losing up to 30 percent of their bone mass during the two-and-a-half year duration of a Mars mission, the implications of which are the cause of some concern among mission planners and medical personnel tasked with ensuring the safe passage of crewmembers.

Table 8.4 Bone loss on Mir (% bone mineral density lost per month) [18]

Variable	Number of Crew Members	Mean Loss (%)	Standard deviation
Spine	18	1.07*	0.63
Neck of femur	18	1.16*	0.85
Trochanter	18	1.58*	0.98
Total body	17	0.35*	0.25
Pelvis	17	1.35*	0.54
Arm	17	0.04*	0.88
Leg	16	0.34*	0.33

*$p < 0.01$

Monitoring bone loss

Key to understanding the mechanisms implicated in bone loss are the means to determine bone mineral demineralization and bone structural measurements. Current methods of measuring BMD include ultrasound, computed tomography, magnetic resonance imaging (MRI) and Dual-Energy Absorptiometry (DEXA). Of these, perhaps the most accurate is DEXA, a system in which two low-dose X-ray beams of different energies are used to scan regions of the body suspected of bone loss. The reason two different X-ray energies are used is to distinguish between bone and muscle since each tissue absorbs differently. Although the results from a DEXA scan provide a reasonably accurate determination of BMD, one of the drawbacks of the system is its inability to distinguish between compact and cancellous bone, making it is almost impossible to reconstruct an engineering model of the bone to perform the necessary stress loading simulations. Since it is necessary to determine the specific location of bone loss to accurately assess fracture risk, a more sensitive means of assessing BMD is required. For Mars missions, this equipment also needs to be flight qualifiable. Such a system is the Advanced Multiple-Projection Dual-energy X-ray Absorptiometry (AMPDXA) system which allows a much higher resolution and precision image to be produced (Figure 8.3).

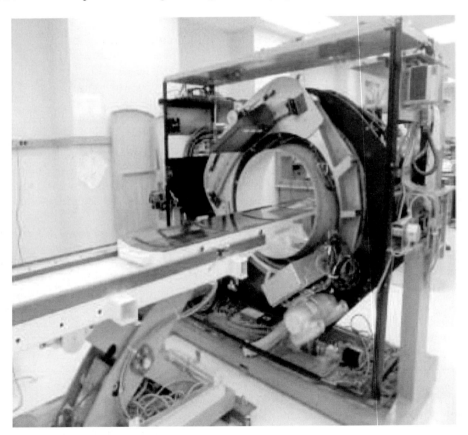

Figure 8.3 The AMPDXA equipment in a clinical setting. A scaled down version of this may be used by astronauts en-route to Mars to assess bone integrity. (AMPDXA for Precision Bone Loss Measurements on Earth and in Space. Harry K. Charles Jr., Michelle H. Chen, Thomas S. Spisz, Thomas J. Beck, Howard S. Feldmesser, Thomas C. Magee, and Barry P. Huang/ Johns Hopkins APL Technical Digest, Volume 25, Number 3 (2004), Page 192.)

The AMPDXA project is being designed by a joint team from APL and the Johns Hopkins School of Medicine. Because the system uses multiple images acquired at different angles, it is possible to determine precise BMD and bone geometry images that may be used for fracture assessment and thereby permit longitudinal studies of bone in space. It is possible a variant of the AMPDXA equipment may be one of the most essential pieces of equipment of a future Mars mission because, by using this system, astronauts and ground-based flight surgeons will be able to accurately monitor rates of bone loss.

While having the ability to monitor bone loss will undoubtedly be helpful, the availability of an AMPDXA system does nothing to reduce the loss of bone. To achieve this, countermeasures must be implemented.

Countermeasures to bone demineralization

The most common countermeasures to bone demineralization may be broadly classified into two categories:

- Pharmacological Intervention.
- Non-pharmacological Intervention.

Pharmacological intervention: osteoporosis drugs

There are a number of osteoporosis drugs approved by the Federal Drug Administration (FDA), the two most recent ones being *alendronate* (marketed under the brand name Fosamax™) and *calcitonin* (marketed under the brand name Miacalcin™). Unfortunately for astronauts and flight surgeons, while the claims made for their efficacy suggest they have potential in bone-loss prevention, there are a number of drawbacks to their use.

Calcitonin, for example, is a hormonal drug resulting in bone mass gains of only one and a half percent a year, a figure far short of the required gains needed to offset losses during a Mars mission. Alendronate, [22] while more effective in promoting bone mass than calcitonin, must be taken for several years to gain the maximum benefit and the side effects of such long-term use are completely unknown. A more controversial drug is Slow Release Sodium Fluoride (SRSF), a formula that boosts the efficiency of bone-building osteoblasts but requires patients to have an annual blood fluoride check to ensure the drug stays below toxic levels in the body!

While the synthetic supplements calcitonin, alendronate and SRSF may not prove effective in offsetting bone loss, a more promising formulation is Osteoporex™, a unique sea-algae calcium that is ninety percent absorbable by the body. Backed by twelve years of research involving more than three hundred treatment studies, the supplement has proved successful in promoting bone mass in ninety-five percent of the studies. An all-natural nutrient supplement that is four times as effective as synthetic pharmaceutical drugs, Osteoporex™ may, in conjunction with other countermeasures described here, prove to be an important part of an effective treatment.

Improved calcium metabolism

Although it may seem one solution to the problem would be to increase calcium intake, the remedy is a little more complex. Simply adding more calcium to the astronaut's diet would not help and may actually exacerbate the problem, because the excessive dietary calcium disrupts the delicate mineral balance needed by the body to repair and build bone. When the body's mineral content is over-weighted in favor of one particular mineral, the vital mineral balance is thrown off, and it becomes more difficult to utilize any of the minerals properly. Because of this, researchers have focused their attention on proper calcium absorption and have discovered calcium balance can be maintained if calcium is used in small doses, in a highly absorbable form, and in proper balance with other absorption-promoting nutrients that enhance calcium metabolism.

Pharmacological intervention: synthetic bone

In addition to administering Osetoporex^{TM} and improving calcium absorption, flight surgeons may suggest astronauts take a product that will grow synthetic bone.

Millenium Biologix Inc. is a medical biotechnology company specializing in the commercialization of innovative products based on synthetic bone technology. Their main area of expertise is the development and application of artificial bone biomaterials that are bioactive. Bioactive means bone cells respond to materials as if they were part of normal bone. In terms of interplanetary astronauts a spaceflight-specific configuration of Millenium's BioBone^{TM} may be the product capable of warding off osteoporosis en-route to Mars since this unique biomaterial has a natural chemical affinity for bone growth proteins and pharmaceutical compounds influential in the bone repair process.

Non-pharmacological intervention

On Earth, non-pharmacological strategies such as exercise combined with adequate calcium, Vitamin D, and protein intake will maintain and even increase bone mass. For example, a daily calcium intake of 0.1–1.0 g/day combined with a Vitamin D intake of 800 IU/day has proven to reduce the risk of fracture. For astronauts travelling to Mars however, more aggressive procedures will be required.

Vitamin D supplementation

Humans need Vitamin D for bone growth and most people achieve their 200 iu/day recommended daily allowance (RDA) from exposure to sunlight. However, Mars-bound astronauts will not be exposed to any sunlight and will be required to maintain their Vitamin D levels by other means.

Studies conducted onboard nuclear submarines suggest in the absence of exposure to sunlight the RDA should be between 500 and 600 iu/day, requiring astronauts to take a multivitamin. Another method Vitamin D levels may be maintained is to provide astronauts with a lighting system containing the required amount of ultraviolet (UV) radiation to stimulate Vitamin D production. Such a simulated sunlight system would have additional beneficial effects on the body by stimulating the production of â-endorphins and promoting a feeling of well-being.

Figure 8.4 Astronauts must exercise between two and three hours per day during long duration missions to offset the effects of bone demineralization. (NASA.)

Exercise

Exercise is one of a variety of countermeasures that has been incorporated into both short and long-duration space flights to help astronauts offset bone loss. During the Mir missions, cosmonauts regularly exercised for between two and three hours a day while being held with strong elastic cords against a running surface and supported by a belt around the waist, a practice that continues onboard the ISS today (Figure 8.4). However, despite various attempts to load the skeleton and all manner of exercise regimes having been employed, crewmembers continue to suffer bone loss and, after four decades of studying the effects of skeletal loading on bone growth, minimum loading thresholds are still unknown.

Low-frequency oscillations

Perhaps one of the more novel ways of maintaining bone density is by standing on a lightly vibrating plate for twenty minutes a day. This approach shifts the focus from trying to stimulate larger loading on bones to creating smaller, high frequency loads by using low-frequency (30-95 Hz) vertical oscillations. The rationale is summed up by a NASA scientist:

> "Currently, most bone researchers believe that stresses placed on bones by e.g. bearing weight or strong physical exertion signals the bone-building cells through some unknown chemical trigger to fortify bones. According to this thinking, the remedy for bone loss in space should be exercise that duplicates

the stresses on our muscles and skeletons experienced during a daily and active life on Earth." [3]

The reason why small, high frequency signals for short periods each time may prove more effective than three to four hour workouts using large loads is unknown. However, research has shown small oscillations such as those arising from muscle contractions that occur when maintaining posture for example, are strong determinants of bone morphology. In one study, groups of female rats were subjected to different stimuli for twenty-eight days to measure the effects on bone formation during microgravity. Microgravity was simulated by suspending some rat groups off the ground by their hind limbs. Some of these groups remained suspended for twenty-eight days, some performed weight-bearing exercise for ten minutes each day, and some were simply subjected to mechanical vibrations for ten minutes a day. The results were encouraging. The group of rats that remained suspended for twenty-eight days experienced significantly decreased bone formation (92%) whereas the group that performed ten minutes of load-bearing every day experienced a less pronounced decrease (61%). The group of rats subjected to mechanical vibrations however experienced a decrease in bone formation of only 3% [3].

Although studies with vibrating plates have not been fully tested on humans, the implications of these studies are exciting and could hold the key to preventing excessive bone loss during a trip to Mars.

Bone loss: a summary

No astronaut has ever gone into space and not experienced a certain degree of bone loss. At one extreme there is the case of NASA astronaut, David Wolf, who spent 128 days onboard Mir and lost up to 12% of his bone mass in certain areas, and at the other extreme is the case of cosmonaut Yuri Romanenko, who spent 326 days in space (MIR EO-3) but did not show any significant bone loss (in fact, Romanenko stood up unaided following the landing and ran 100 meters the next day!). Equally puzzling is the variation in recovery times between astronauts. Some astronauts experience bone loss for a period after the mission before reversal, whereas others do not experience reversal for several years. Based on these idiosyncrasies, one means of selecting Mars astronauts may be to select based on a genetic predisposition to bone loss, but unfortunately no means to test for this exists. Another means might be simply to select those astronauts with the greatest bone density since these candidates would simply have more to lose, as alluded to in the article "Running to Mars" [24]. Yet another method of mitigating bone loss would be to create artificial gravity but this would only provide a fraction of the pull of Earth's gravity. Exercise is obviously important, and in the absence of a breakthrough in bone loss countermeasures, it can be expected Mars astronauts will exercise three to four hours a day. However, since no single countermeasure mentioned will solve the bone loss problem, the most likely scenario is a combination of several countermeasures. Perhaps exercise combined with drug therapy and vibration therapy may prove the most effective, or maybe low-frequency vibration regime and exercise combined may do the trick. The problem researchers have is determining which combination will be the most effective.

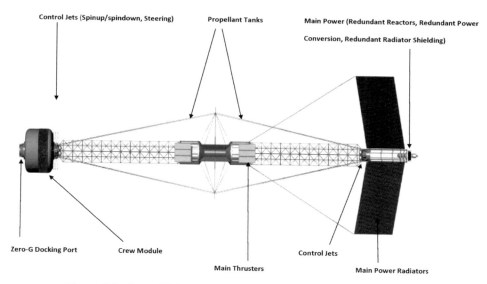

Figure 8.5 An artificial gravity vehicle proposed by NASA. (NASA.)

MUSCLE LOSS

As bone mass is lost in zero gravity, muscle undergoes a similar process, as evidenced by dozens of studies documenting changes to muscle mass and architecture during long-duration spaceflight [13, 23]. During their journey to and from Mars, the absence of mechanical strain on the lower limbs of the astronauts will result in muscles used for posture being most affected. Gradually, in spite of a vigorous exercise regime, the crew will lose strength and size in the muscles used for standing, walking and posture as a result of a process known as muscle atrophy.

The first indication of how atrophied their muscles are will be shortly after the crew lands on Mars. Because the muscles of their lower limbs will have wasted away, astronauts will exhibit a variety of postural control and coordination problems, forcing them to adopt unusual strategies to perform even the most basic of movements. In fact, it may take the crew some time just to perceive the degree to which their postural control capabilities are compromised. Consequently, astronauts conducting surface operations will be more susceptible to potentially disabling ligament tears and tendon ruptures for which clinical experience indicates complicated and unpredictable healing patterns. Furthermore, crewmembers with atrophied muscles will be unable to perform EVA scenarios encompassing the full spectrum of activities and may find their endurance capability severely compromised.

Unfortunately, in common with the mechanism of bone loss, the process of muscle atrophy is poorly understood and to date there are no effective non-exercise countermeasures, which means crewmembers must spend two to three hours every day either on the treadmill or on the exercise bike. However, promising research is being conducted in the area of artificial gravity generated by rotation of the entire space vehicle or of an inner chamber. Artificial gravity (Figure 8.5) is the centripetal force generated in a rotating vehicle and is proportional to the product of the square of angular velocity and the radius of rotation. For a particular G-level, there is a trade off between velocity of rotation and radius and since increased radius is more expensive to achieve than velocity, most research is directed at defining the highest rotation rate to which humans can adapt.

CARDIOVASCULAR CHANGES

Prolonged exposure to zero gravity and reduced gravity cause the cardiovascular system to undergo adaptive changes in both function and structure. These changes result in cardiovascular deconditioning, a problem that grows in severity with increased spaceflight duration. While cardiovascular deconditioning has not presented a serious inflight issue, astronauts preparing to set foot on Mars may experience problems adapting to Martian gravity due to the debilitating effects of orthostatic intolerance and cardiac dysfunction.

Orthostatic intolerance
After spending four or more months in the zero gravity environment of deep space, astronauts landing on Mars will experience orthostatic intolerance caused by a dramatic drop in blood pressure. The condition will begin to manifest itself as soon as the vehicle enters the gravitational influence of Mars, at which point blood will begin to pool in the lower extremities, causing a reduction in the volume of blood in the chest and head area. The precipitous drop in blood pressure that follows will result in crewmembers feeling dizzy and discombobulated to the degree that some of them may even faint. Of course there are countermeasures to orthostatic intolerance, including saline-loading which boosts the blood volume, thereby reducing the severity of the symptoms. Astronauts can also increase the pressure on the cuffs of their G-suits, but even this measure combined with saline-loading may be insufficient to prevent crewmembers from becoming orthostatically intolerant.

Cardiac Dysfunction
Due to the prolonged time spent in the absence of gravity, astronauts will lose significant cardiac mass even if they maintain a rigorous exercise regime. Whether such cardiac changes will be reversible is unknown but perhaps of more concern to the astronauts and flight surgeons is the possibility that such prolonged cardiovascular deconditioning may exacerbate previously undetected heart problems such as abnormal heart rhythms.

NEUROVESTIBULAR

Space Adaptation Syndrome

The most pronounced effect of entering zero gravity is Space Adaptation Syndrome (SAS), caused by the vestibular system's inability to distinguish direction. Often known as space motion sickness, the syndrome is usually characterized by nausea and vomiting but symptoms usually resolve themselves within three days. While this may not be a problem for crews after the first three days of the mission, a similar syndrome is experienced when re-adapting to gravity, resulting in performance deficits, altered visual and motion cues and even a certain level of discombobulation.

Sensorimotor and locomotion adaptation

While SAS resolves quickly, other components of the neurovestibular system require months to adapt in flight, and several days to readapt to gravity. A good example of this readaptation was provided by the Apollo astronauts who regularly fell while experiencing locomotion problems during their lunar EVAs. After several months confined to a vehicle the size of the proverbial school bus, it isn't difficult to imagine the balance and locomotion problems caused by donning a bulky spacesuit and trying to adapt to gravity, albeit reduced.

IMMUNOLOGICAL

Ever since human spaceflight began, it has been known astronauts are susceptible to illness as evidenced by the fact fifteen of the twenty-seven Apollo astronauts contracted bacterial or viral infections either during their missions or within a week of returning to Earth. Why spaceflight alters immune function is still a mystery to scientists but it is known spaceflight affects the activation of T-cells, a type of white blood cell that defends the body against disease. In one study, Millie Hughes-Fulford, an ex-NASA astronaut turned medical professor at the University of California, subjected human immune cells to simulated zero gravity. Normally, when the body detects an unfamiliar entity such as a virus, a signalling system known as the Protein Kinase A (PKA) pathway responds by activating ninety-nine genes which in turn galvanize the T-cells into action. Once activated, the T-cells set about destroying the virus but in the simulated zero gravity condition of the study, ninety-one of the genes simply didn't activate, an indication to Hughes-Fulford that space short-circuits much of the body's immune response. This shouldn't really come as a surprise given that humans evolved in the Earth's gravity field, but a suppressed immune system is a serious concern for flight surgeons and crewmembers, especially when the exacerbating effect of radiation is considered. Unsurprisingly, radiation makes matters even worse for the depressed immune system.

BEHAVIORAL ISSUES

The history of polar exploration is rife with tales of heroism, self-sacrifice and conquest. It also has many tales of hardship, suffering, illness and death. Thanks to meticulous diaries maintained by explorers such as Nansen and Shackleton, today's human performance scientists have a good understanding of the demands people on long-duration expeditions face from the harsh physical and social environment (Table 8.5) of long-duration spaceflight. This understanding is of great importance for the prevention and treatment of mortality related to poor neurobehavioral and psychosocial adaptation to the extreme and unusual environment of interplanetary space, the details of which are discussed here.

Table 8.5 Behavioral stressors of long duration spaceflight

Psychological	Psychosocial	Human Factors	Habitability
Isolation & Confinement	Team coordination demands	High & low work-loads	Limited hygiene
Limited abort options	Interpersonal tension	Limited communication	Chronic noise exposure
High risk condition	Family life disruption	Limited equipment	Limited sleep facilities
Mission complexity	Enforced interpersonal contact.	Mission danger Risk of equipment malfunction	Lighting and illumination
Hostile environment	Crew factors (Size, gender)	Adaptation to environment	Lack of privacy
Sensory stimuli altered	Multicultural issues	Food restrictions	Isolation from support

Expedition stressors

The psychosocial environment of the Mars expedition will be characterized by isolation and confinement. During the transit to and from Mars, crewmembers will be physically isolated from Earth, while on the surface of the Red Planet they will be confined to a habitat the size of a school bus surrounded by the most hostile environment faced by humans to date. Despite rigorous selection procedures, it is natural that crewmembers separated from family and friends will feel degrees of emotional deprivation. Social confinement will be exacerbated by absence of privacy, and little separation will exist between work and leisure because the living and working spaces will be so close to one another and each crewmember will interact with the same group of individuals in both types of activities. To better understand the psychological changes crewmembers may experience during a Mars mission, it is helpful to categorize the effects into behavioral problems, psychiatric disorders and positive effects.

Behavioural problems

Interpersonal tension and conflict

"All the conditions necessary for murder are met if you shut two men in a cabin measuring eighteen feet by twenty and leave them together for two months."

Valery Ryumin, veteran cosmonaut.

As Ryumin suggests, sustained, close personal contact with other crewmembers may be extremely stressful, a situation compounded by the stressors of isolation, dangers posed by radiation, equipment malfunctions and boredom. Polar explorers faced with interpersonal problems had the option of removing themselves temporarily from the source of stress by simply going for a walk. Unfortunately, interplanetary explorers, confined to a living space the size of a small motor-home, shared with three to five other crewmembers, will not have this option. Inevitably, during the cruise phase of the voyage, each crewmember's repertoire of jokes, personal experiences and anecdotes will become increasingly familiar. Mannerisms, which were initially innocuous, will be exaggerated in the minds of a crew subject to the constraints of their extended confinement. Eventually, the most minor irritation will assume unreasonable proportions and will force some crewmembers to retreat. Unfortunately, the only escape will be the lavatory or the small coffin-sized compartment that serves as a sleep cubicle.

Once again, polar exploration provides examples, both good and bad, of how interpersonal problems may affect an expedition. Perhaps the most successful examples include Shackleton's Imperial Trans-Antarctic Expedition and Nansen's Polar Expedition which consisted of a group of thirteen Norwegians who survived three years in polar isolation, enduring conditions making any Mars expedition a five star luxury cruise by comparison. Both Shackleton's and Nansen's expeditions experienced interpersonal problems but, thanks to rigorous selection criteria, disruption was kept to a minimum. Undoubtedly, successful Mars voyages will be those missions adopting the expeditionary approach of Shackleton and Nansen.

Psychological closing, autonomization and displacement

The phenomenon of *psychological closing* is manifested by decreased communication intensity as well as increased filtration of the scope and content of crew communication. For example, it has been observed during space missions that crewmembers tend to conceal medical and psychological problems and also demonstrate preferences in contacts with certain mission control personnel [7].

Autonomization is expressed by crew egocentrism and is considered a natural and even necessary stage of the formation of cohesive group of isolated and confined individuals. The problem with this phenomenon occurs when it is manifested by mission control personnel being perceived as opponents and not partners. In such a situation, the crew becomes critical in their discussions with mission control, resulting in the compromise of operational effectiveness. A good example of the phenomenon of autonomization was the crew of Skylab who complained to mission control about being overworked [11].

Closely related to autonomization is the phenomenon of *displacement*, which occurs when crewmembers experience high levels of anxiety and interpersonal conflict that cannot be resolved directly. In such a situation, the unpleasant effects may sometimes be externalized to mission control personnel. The cause of the displacement is a coping strategy, allowing crewmembers to avoid open conflict by venting their frustrations on the unfortunate mission control personnel. Although the strategy works for a while, in the long term it encourages negative feelings, territorial behavior and a disintegration of group cohesion.

Boredom

> "Oh! At times this inactivity crushes one's very soul; one's life seems as dark as the winter night outside... I feel I must break through this deadness, this inertia, and find some outlet for my energies. Can't something happen?
> *Dr. Fridtjof Nansen, legendary Norwegian polar explorer, 1897.*

In sharp contrast with the tight schedule and high workloads characterizing a Space Shuttle mission or an ISS increment, the task requirements of a Mars-bound astronaut will be almost non-existent. Between the high-tempo flurries of activity following departure from Earth orbit and preparation for arrival at Mars, crewmembers will be required to perform relatively few meaningful tasks beyond the scheduled systems checks, emergency abort rehearsals and preparations for landing. Consequently, the cruise phase to and from Mars will be characterized by the inevitable behavioral consequences of low-workload conditions, inactivity and boredom experienced by many polar explorers and Antarctic winter-over personnel.

Fortunately, for mission planners anxious about the effects of tedium and lethargy on Mars crews, there is an abundance of knowledge gleaned from polar expeditions. Perhaps one of the best examples of how crews staved off the demoralizing effects of boredom is the case of Ernest Shackleton's Imperial Trans-Antarctic Expedition in 1914.

Shackleton's objective was to cross the Antarctic continent, but when he and his crew were still hundreds of kilometers from the intended base, his ship, the *Endurance*, first became trapped and was then crushed by pack ice. Shackleton and his crew abandoned the *Endurance* before she sank and survived for months on ice floes and, later, on a desolate island. Despite their desperate situation, the crew kept themselves busy packing rescue supplies, practicing escape procedures, and searching for food. However, after several months of preparation and practice, the crew simply ran out of things to do.

> "Time, indeed, was beginning to weigh a little heavily. Each day blurred anonymously into the one before. Though they invariably tried to see the good side of things, they were unable to fight off a growing sense of disappointment."
> *Crewmember of Ernest Shackleton's Imperial Trans-Antarctic Expedition.*

As the pack ice disintegrated around them, Shackleton and his crew made a break for the open sea in a desperate attempt to reach Elephant Island. There, the crew planned a last-ditch effort to reach South Georgia Island, more than one thousand kilometers distant. While Shackleton and five crewmembers made the journey, the remaining twenty-two crewmembers waited on Elephant Island with little to do but wait and talk about the possibility of rescue.

Studies of the anecdotal accounts of polar expedition crewmembers have revealed meaningful, non-repetitious work is the most effective means to pass time and stave off lethargy and boredom. Such countermeasures are currently employed by Antarctic program managers and it is probable that the implementation of such a goal-oriented approach will be implemented into the daily flight schedule of those travelling to Mars.

Psychiatric disorders

Despite rigorous selection criteria, psychiatric events have occurred during space missions. Crewmembers on a Mars mission will be exposed to a much higher level of autonomy and long-term confinement and isolation than any previous space crew. Some psychologists have suggested this may increase the psychological risks related to individual crewmember performance and will produce new psychological challenges never before experienced. In the context of space travel, psychiatric issues are abnormal psychological and interpersonal responses to the conditions of the spaceflight environment. Such issues may develop due to operational mission stressors, such as confinement and radiation, and psychosocial factors such as crew tension and cultural differences.

Asthenia is a common psychiatric condition that may affect astronauts. The syndrome is defined as a weakness of the nervous system, resulting in fatigue, irritability, concentration difficulties, restlessness and physical weakness. In the Russian space program, *asthenization* is carefully monitored and countermeasures are employed to prevent it, while in the United States, the syndrome is addressed from a different perspective, resulting in a NASA flight surgeons possibly under-detecting asthenic distress.

Although a psychiatric event among the crew is an unlikely occurrence, there will be procedures in place in the event of such a contingency. For example, it is anticipated at least two of the crewmembers will be trained in counseling, crisis intervention techniques, and the administration of psychoactive medications such as tranquilizers. In the event an astronaut becomes potentially suicidal or violent, facilities will be available to seclude or restrain the affected crewmember.

Positive effects

While it is inevitable crewmembers will experience some of the aforementioned symptoms and disorders in varying degrees during their two-to-three-year mission, it is also certain they will experience several positive effects (Table 8.6). Some astronauts have even reported transcendental experiences and a better sense of the unity of mankind as a result of viewing the Earth below and the cosmos beyond [4]. Despite the risks and the acknowledged dangers, it can be anticipated crewmembers

Table 8.6 Positive characteristics of a manned Mars mission

Excitement of encountering and learning about unknown and unusual terrain.

Challenges of expedition tasks and environment.

Camaraderie and mutual support from other crewmembers and mission control.

Resilience and coping while performing mission tasks in hostile and challenging environment.

Increased fortitude, perseverance, ingenuity as a result of meeting the expedition challenges.

will be positively affected by the grandeur of their surroundings and the thrill of overcoming the challenges of travelling to and living on Mars. Other positive effects will surely be generated by the inherently enjoyable characteristics of the expedition experience, although it is important to note that positive and negative reactions are not mutually exclusive. For example, people can enjoy and benefit from an experience even though they may also be in discomfort.

Planetary missions, in common with the expeditions of Nansen, Shackleton and Amundsen, will always be inherently risky. Due to the unpredictable nature of the interplanetary space environment, perhaps the greatest dangers are those affecting human health and performance. Clearly, given the medical unknowns concerning the quality, intensity and effects of radiation and the deleterious effects of bone demineralization, undertaking a manned Mars mission will provide a daunting challenge to the medical profession tasked with protecting crewmembers from the hostile environment of interplanetary space and providing the best medical care possible during these remote journeys. It will be the task of the physicians to estimate and explain the medical risks, so NASA, mission planners and the crew can weigh them against the overall risks and benefits of the mission, before deciding whether to accept them. There is no definite threshold short of which we are not ready, and beyond which we are ready, to embark on a Mars mission. We will always be ready if the risks are judged worth taking.

The risks and unknowns faced by Nansen, Shackleton and Amundsen and their crews were of similar magnitude to those faced by future Mars explorers, yet these great polar explorers were still willing to spend years away on expeditions characterized by horrific cold, limited supplies, and tuberculosis. No shirt-sleeve environment of a pressurized Mars base for them. Why, knowing the risks, did these men embark on such expeditions? Some went along for the money, and in part, for the fame, but mostly because the urge to explore was embedded so deep in them that the dark spectre of death was no deterrent. The point is, despite all the risks discussed here, humans are more than capable of surviving and staying sane on multi-year missions in horrendous conditions and in the face of appalling hardship. Nansen, Shackleton and Amundsen endured conditions that would make even the strongest men quail, and yet they survived. Their stories and exploits provide the rest of us with a benchmark against which all other human endeavors appear downright easy, and so it will be with the first expedition to Mars. Chapter 9 describes how this expedition will unfold.

REFERENCES

1. Anderson, B.M.; Clowdsley, M.S.; Qualls, G.D.; Nealy, J.E. Nuclear Radiation Fields on the Mars Surface: Risk Analysis for Long-Term Living Environment. SAE International Conference on Environmental Systems (ICES) Paper 2005-01-2822. 2005.
2. Banfi, A.; Bianchi, G.; Galotto, M.; Cancedda, R., and Quarto, R. Bone marrow stromal damage after chemo/radiotherapy: occurrence, consequences and possibilities of treatment. Leuk Lymphoma, 42, pp. 863–870, 2001.
3. Barry, P.L. Good Vibrations: A new treatment under study by NASA-funded doctors could reverse bone loss experienced by astronauts in space. Science@NASA. 4 May, 2002.
4. Connors, M.M.; Harrison, A.A.; Akins, F.R. Living Aloft: Human Requirements for Extended Spaceflight, NASA SP-483, NASA, Washington, DC, 1985.
5. Damewood, M.D., and Grochow, L.B. Prospects for fertility after chemotherapy or radiation for neoplastic disease. Fertility Sterility. 45: 443–459. 1986.
6. Donovan, D.J.; Huynh, T.V.; Purdon, E.B.; Johnson, R.E., and Sniezek, J.C. Osteoradionecrosis of the cervical pine resulting from radiography for primary head and neck malignancies: operative and nonoperative management. Case report. J. Neurosurg Spine 3, pp. 159–164, 2005.
7. Gushin, V.I.; Yusupova, A.; Popova, I. Crew-ground control communication styles: preliminary results, IAC-04-G.5.a.06, 55th International Astronautical Congress. Vancouver, B.C., Canada, October 4–8, 2004.
8. Hamilton, S,A.; Pecaut, M.J.; Gridley, D.S,; Travis, N.D.; Bandstra, E.R.; Willey, J.S.; Nelson, G.A., and Bateman, T.A. A murine model for bone loss from therapeutic and space-relevant sources of radiation. J. Appl. Physiol. 101: pp. 789–793. 2006.
9. Hada, M. and Sutherland, B.M. Spectrum of Complex DNA Damages Depends on the Incident Radiation. Radiation Research. 165, 223. 2006. doi:10.1667/RR3498.1.
10. Hahn, E.W.; Feingold, S.M.; Simpson, L.; Batata, M. Recovery from aspermia induced by low-dose radiation in seminoma patients. Cancer. 50: 337–340. 1982.
11. Kelly, A.D.; Kanas, N. Communication between space crews and ground personnel: a survey of astronauts and cosmonauts. Aviation, Space and Environmental Medicine. 64. 795–800. 1993.
12. Kopjar, N.; Mioèiæ, S.; Ramiæ, S.; Miliæ, M., and Viculin, T. Assessment of the Radioprotective Effects of Amifostine and Melatonin on Human Lymphocytes Irradiated with Gamma-rays In Vitro. Arh. Hig. Rada. Toksikol. 57: 155–163. 2006.
13. LeBlanc, A.; Rowe, R.; Schneider, V.; Evans, H., and Hedrick, T. Regional Muscle Loss after Short Duration Space Flight. Aviation Space Environmental Medicine. 26, pp. 20–24, 1992.
14. LeBlanc, A.; Schneider, V.; Shackelford, L.; West, S.; Ogavov, V.; Bakulin, A., and Veronin, L. Bone mineral and lean tissue loss after long duration spaceflight. Journal of Bone Mineral Research 11, pp. S323, 1996.
15. Levy, R.H., and Janes, G.S. Plasma radiation shielding for deep space vehicles. Space/Aeronautics. 45: 106–120. 1966.
16. Lushbaugh, C.C., and Cassarett, G.W. Effects of gonadal irradiation in clinical radiation therapy: A Review. Cancer 37: 1111–1125. 1976.

17. Meistrich, M.L. and Van Beck, M.E.A.B. Radiation sensitivity of the human testes. Advanced Radiation Biology. 14: 227–268. 1990.

18. Oganov, V.S.; Grigoriev, A.I.; Voronine, L.I.; Rakhmanov, A.S.; Bakulin, A.B.; Schneider, V., and LeBlanc, A. Mineral density of bone tissue in cosmonauts after 4.5–6 month missions on MIR. Kosmicheskaya Biologiya I. Aviakosmischeskaya Meditsina, Vol. 26, No. 5, pp. 20–24, (in Russian) 1992.

19. Raju, M.R.; Bain, E.; Carpenter, S.G.; Cox, R.A., and Robertson, J.B. A heavy particle comparative study. Part II: Cell survival versus depth. British Journal of Radiology. 51: 704–711. 1978.

20. Rapp, D. Radiation Effects and Shielding Requirements in Human Missions to the Moon and Mars. The International Journal of Mars Science and Exploration. Pp46-71. 2006. do:10.1555/mars.2006.0004.

21. Rubin, C. ; Xu, G. ; Judex, S. The Anabolic Activity of Bone Tissue, Suppressed by Disuse, is Normalized by Brief Exposure to Extremely Low-Magnitude Mechanical Stimuli. The FASEB Journal. 15. pp: 2225–2229. 2001.

22. Ruml, L.A.; Dubois, S.K.; Roberts, M.L., and Pak, C.Y.C. 1995. Prevention of hypercalcuria and stone-forming propensity during prolonged bedrest by alendronate. *Journal of Bone Mineral Research*, 10, pp. 655–662, 1995.

23. Sawin, C.F.; Taylor, G.R.; Smith, W.L. Eds. Extended Duration Orbiter Medical Project. Final Report 1989-1995. NASA/SP-534. Johnson Space Center, Houston, TX: National Aeronautics and Space Administration. 1999.

24. Seedhouse, E. Running to Mars. Spaceflight. Vol. 39. pp. 266–267. 1997.

25. Setlow, R.; Dicello, J.F.; Fry, R.J.M.; Little, J.B.; Preston, R.J.; Smathers, J.B., and Ullrich, R.L. Radiation Hazards to Crews of Interplanetary Missions: Biological Issues and Research Strategies. Washington, DC: Space Studies Board, National Research Council, & National Academy Press, pp. 13–34, 1996.

26. Simonsen, L.C., and Nealy, J.E. Mars Surface Radiation Exposure for Solar Maximum Conditions and 1989 Solar Proton Events. NASA Technical paper 3300. February 1993.

27. Todorov, S.L.; Grigoriev, Y.G.; Rizhov, N.I.; Ivanov, B.A.; Malyutina, T.S., and Micleva, M.S. Dose response relationship for chromosomal aberrations induced by X-rays or 50 MeV protons in human peripheral lymphocytes. Mutat. Res. 15: 215. 1972.

28. Townsend, L.W. Galactic heavy-ion shielding using electrostatic fields. NASA Technical Memorandum 86265. September, 1984.

29. Vernikos, J., and Nicogossian, A.E. Strategic Program Plan for Space Radiation Health Research. Washington, DC: NASA Headquarters Space Radiation Health, pp. 1–71, 1998.

30. Vico, L.; Collet, P.; Guignandon, A.; Lafarge-Proust, Marie-Helene.; Thomas, T.; Rehailia, M.; Alexandre, C. Effects of Long-term Microgravity Exposure on Cancellous and Cortical Weight-bearing Bones of Cosmonauts. The Lancet. 355. pp: 1607–1611. 2000.

31. Vogel, J.M., and Whittle, M.W. Bone Mineral Content Changes in the Skylab Astronauts. In Proc. Am. Soc. Roentgenol. 126, pp. 1296–1297, 1976.

32. Wilson, J.W.; Nealy, J.E., and Schimmerling, W. Effects of Radiobiological Uncertainty on Shield Design for a 60-Day Lunar Mission." NASA Technical Memorandum 4422. 1993.

33. Wu, B.; Medvedovsky, C., and Worgul, B.V. Non-subjective cataract analysis and its application in space radiation risk assessment. Advances in Space Research. 14: 493–500. 1994.

9

Voyage to Mars

"Some men see things as they are and say, 'Why?' I dream of things that never were and say, 'Why not?'"

Robert F. Kennedy. 1968 Presidential Campaign

A journey to Mars will rival the journeys of Shackleton, Amundsen and Nansen, especially in terms of mission duration and isolation from sources of supply and assistance. Once in transit from Earth to Mars, the crew will have no re-supply and only limited resources and capabilities available to them to maintain and repair the vehicle. The International Space Station (ISS) is outfitted with a lifeboat to assure crew safety if a system failure occurs or astronaut illness warrants a return to Earth, but astronauts en-route to Mars will have no such luxury. Once the vehicle performs its trans-Mars insertion (TMI) burn, the crew is committed to a trajectory that will take them to the Red Planet, and support from Earth will be limited to communications and ground-based experience. But what exactly will a manned mission to Mars involve from the astronauts perspective? How will they live onboard the cramped spacecraft for four or more months? What will it be like to be isolated for several months, with little chance of return in the event of a malfunction? The answers to these questions form the content of this chapter which describes how a manned Mars mission will unfold, phase by phase (Table 9.1).

Table 9.1 Phases of a manned mission to Mars

Phase	Description	Phase	Description
1	Pre-launch Activities	7	Crewed Long-term Surface Operations
2	Launch & Low Earth Orbit Operations	8	Departure Preparations & Departure
3	Trans-Mars Injection & Interplanetary Travel	9	Rendezvous, Docking & Transfer to Earth Return Vehicle
4	Arrival at Mars & Orbit Capture	10	Trans-Earth Injection & Inter-planetary Travel
5	Mars Entry, Descent & Landing	11	Earth, Entry & Landing
6	Crewed Initial Surface Operations	12	Post Landing Recovery

MISSION RISK

Before describing the nuts and bolts of the first interplanetary voyage, it is important to understand the risks faced by those chosen to go to Mars. While the biomedical and behavioral challenges described in the preceding chapter may seem ominous enough, the hazards posed by system or multiple system failure are by several orders of magnitude more dangerous. Potentially, there are thousands of failure points in the vehicle and mission architecture, some of which may result in loss of crew (LOC) or loss of mission (LOM). Although several critical systems will be designed with one or more levels of redundancy, some systems, such as those supporting entry, descent and landing (EDL), will be, by necessity single-string designs. As with all manned missions preceding it, a Mars expedition crew will be trained exhaustively for all possible contingencies mission planners can foresee. An example of how LOC, LOM and non-catastrophic events may be categorized is provided in Table 9.2, while Tables 9.3 (a)-(f) describe some of the emergencies that may occur during a transit to and from Mars.

Table 9.2 Criticality codes

Loss of Crew (LOC) Criticality Codes

Code	Failure
1	Failure may cause LOC
1H	Failure may cause LOC or loss of vehicle, but redundant hardware is available.
1A	Failure of systems/subsystems endangering crew, but survival possible using alternate solution. No redundant hardware available.

Loss of Mission (LOM) Criticality Codes

Code	Failure
2	Failure may cause LOM.
2H	Failure may cause LOM, but redundant hardware is available.
2A	Failure of systems/subsystems endangering vehicle, but continuation possible using alternate solution. No redundant hardware available.

Non-Catastrophic Criticality Codes

Code	Failure
3	Failure that may cause moderate damage without risk of LOM or LOC.
3H	Failure that may cause moderate damage but redundant hardware available.
3A	Damage to systems/subsystems vehicle, but continuation possible using alternate solution. No redundant hardware available.

Tables 9.3 (a)–(f) Systems failures travelling to and from Mars

Table 9.3 (a) Structural Integrity Failures

Failure	Code	Consequences	Action
Damaged Aerobrake	2	LOM	Return to Earth
Small hole in MTV	2-3	Potential LOM	Conduct EVA to repair hole. If unsuccessful, return to Earth.
Large hole in MTV	1-1A	Potential LOC and LOM	If hole occurs in interplanetary space, LOC likely.
Failure of MTV/MHV separation mechanism.	2-2A	Potential LOM.	Crew cannot land if MHV does not separate from MTV. Contingency EVA will be performed.

Table 9.3 (b) Soft Landing System Failures

Failure	Code	Consequences	Action
Aerobrake fails to function as heatshield on entry into Martian atmosphere.	1	LOC.	If aerobrake fails, vehicle will burn up on entry and crew will be lost.
Aerobrake jettison mechanism failure.	1	LOC	If aerobrake fails to release from MHV after Mars entry phase, craft will be unable to land properly on surface and will likely crash.
Primary parachute failure.	1H	Potential LOC	Back-up parachutes used.
Failure of secondary parachutes.	1H	Potential LOC.	Landing can be attempted using landing thrusters only since extra fuel is allocated for this purpose but success unlikely.
Landing thruster failure (1)	1H	Potential LOC.	Redundant thrusters used.
Landing thruster failure (all)	1	LOC	Vehicle will be unable to slow down in final phase of descent and will crash.

Table 9.3 (c) Power System and Communications Failures

Failure	Code	Consequences	Action
Loss of wattage	1A, 2A, 3A	Loss of system(s) and subsystem(s)	Power system is modular in design and features redundant systems. Time-sharing & power management strategies implemented.
High-gain antenna failure.	3H	Temporarily limit data return to Earth.	Use back-up antenna.
Secondary antenna failure.	3H	Data return to Earth limited.	Vehicle will operate autonomously. Crew will use low-gain antenna.
Primary low-gain antenna failure.	3H	Temporarily limit data return to Earth.	Use secondary antenna.
Secondary low-gain antenna failure.	3H	No data return to Earth.	Crew and vehicle operate autonomously in accordance with contingency procedures.

Table 9.3 (d) Life Support Failures

Failure	Code	Consequences	Action
Partial failure of water recycling system.	1A	Temporary loss of clean water.	Implement redundant/ back-up hardware.
Total failure of water recycling system.	1H	Potential LOC	Crew will attempt repair. If unsuccessful, LOC.
Partial failure of air pressurization system.	1H	Loss of primary air pressurization system.	Switch to redundant/back-up hardware.
Total failure of air pressurization system.	1	Potential LOC	Crew will attempt repair. If unsuccessful, LOC.
Partial failure of air management system.	1H	Loss of primary air management system.	Switch to redundant/back-up hardware.
Total failure of air management system.	1	Potential LOC.	Crew will attempt repair. If unsuccessful, LOC.

Table 9.3 (e) Main Engine, Fuel Tank, OMS and RCS Failure

Failure	Code	Consequences	Action
Main engines do not ignite	2	LOM	If one of more main engines does not ignite, vehicle cannot depart Earth orbit.
Single engine failure during trans-Mars injection burn.	2H	Temporary loss of thrust.	Remaining engines increase thrust to compensate for lost engine.
One of more engines catches fire.	1-1H	Potential LOM and/or LOC.	If this occurs in LEO, the crew separate from MHV and rendezvous and dock with the ISS. If this occurs in transit, LOC.
Failure of primary valve to open.	2H	Temporary delay to starting engine.	Switch to redundant/back-up system.
Failure of secondary valve to open.	2H	Temporary delay to starting engine.	Switch to back-up fuel tanks.
Multiple tanks fail to open.	2	Delay to starting engine.	Consider contingency EVA. If unsuccessful, LOM.
Fuel tank explodes	1	LOC	Explosion would likely destroy vehicle and kill crew.
Failure of Orbital Manoeuvering System in LEO.	2R	Mission delayed.	Use secondary engines to rendezvous and dock with MHV. If OMS fails completely, abort mission.
Total failure of OMS in transit	2H	Mid-course corrections not possible.	Attempt mid-course corrections with RCS. Consider mission abort.
Total failure of OMS in Mars orbit.	3H	Orbital corrections and rendezvous and docking capability impaired.	Attempt use of RCS to effect rendezvous and docking. Consider mission abort.
Primary RCS failure	3H	Temporary loss of RCS.	Switch to secondary RCS.
Secondary RCS failure.	3H	LOM	Consider contingency EVA to repair. If unsuccessful, LOM.

Table 9.3 (f) Video, Computer and Miscellaneous Systems Failure

Failure	Code	Consequences	Action
Single or complete internal camera failure	3H	Temporary loss of video feed.	Crew repair system or replace unit with new one from storage.
Failure of primary internal bus system.	3H	Temporary loss of video feed.	Switch to secondary bus system.
Failure of secondary internal bus system.	3H	No return of images.	Attempt repair. If unsuccessful, mission impaired.
Monitor failure.	3H	Temporary loss of data.	Switch to secondary system.
Transmitter/receiver failure (Ground)	3H	Temporary loss of data.	Switch to secondary system.
Guidance & Navigation Computer Failure	3H	Temporary loss of navigation & guidance.	Voting system votes out faulty computers which are replaced in accordance with redundancy architecture.
Failure of aerostationary satellite (while on Mars surface).	3H	Communication with Earth impaired.	Use surface communication network to relay via orbiting MTV.

Given the number of things that can go wrong, it would be unnatural for the astronauts not to become preoccupied with the dangers but, as the launch date approaches, the activities requiring their attention will probably distract them, at least for while. Some of these activities are described here in an account of a generic mission following the SpaceWorks Inc. (SEI) architecture, described in Chapter 3.

PRE-LAUNCH ACTIVITIES

As the launch approaches, training intensity begins to slacken and, a week before launch, the crew moves into the Astronaut Quarantine Facility (AQF) in Houston (Figure 9.1). Subsequently, the crew quarters at Kennedy Space Center (KSC) where nobody, not even their families, is allowed entry, unless screened by the flight surgeon. A couple of days before launch the three astronauts meet their families for the final time, separated by a one-meter ditch on a patch of grass close to the crew quarters.

Figure 9.1 Astronaut Quarantine Facility in Houston. (NASA.)

LAUNCH AND INITIAL LOW EARTH ORBIT OPERATIONS

On launch day, the crew has breakfast, followed by the standard crew photo and a weather briefing, before donning launch and entry suits. After the photo, the astronauts engage in one more ritual before catching the bus that will take them to the launch pad. Suited up in their orange pressure suits, the astronauts play a card game called Possum's Fargo, a fighter pilot game where the worst hand wins. They keep playing until the commander loses his hand, after which they board the bus and depart for Launch Pad 39-B.

Three hours before launch, the crew finally faces the Ares I/Orion stack from the gantry of the KSC launch complex (Figure 9.2), almost sixty meters above the ground. With the aid of United Space Alliance (USA) technicians dressed in white, the crew crawls into Orion and checks communications. Like parents fussing over their kids, the technicians help the astronauts into their seats, making sure straps are tightened, helmets adjusted and visors are up. Meanwhile, the countdown continues as the commander switches the event timer to start, confirms the operations recorders are on, and scans the instrument panel.

Outside, as launch events unfold, roadside Mars Mission parties are in full swing along the NASA causeway, and a million spectators crowd the area from KSC to Titusville waiting to see the launch of the world's first interplanetary mission. At the KSC Visitors Complex, spectators and press listen to NASA's public affairs officer (PAO) explain the stages of the countdown as the capsule communicator (CAPCOM) booms across the VIP stand, running down a series of standard NASA acronyms.

At launch minus two minutes, the crew pull their visors down and hear the familiar commands from mission control informing them the redundant launch control has commenced. Twenty-five seconds from launch, the general purpose

Figure 9.2   Ares I/Orion on the launch pad. (NASA.)

computers (GPCs) take control of the launch sequence and at fifteen seconds, the commander checks the pyrotechnics are armed and the sound suppression system activated. At T minus eight seconds, the position vector of Ares I is loaded, meaning the geographic location of the vehicle has been turned into positional data inside the GPCs. At T minus one second, the crew waits for confirmation that engine status lights are green and SRB ignition. As the SRB ignites, the spectators see a plume of dazzlingly yellow light from the SRB and a brief flash as pyrotechnics sever the hold-down bolts securing the vehicle to the pad. As Ares I lifts off the ground (Figure 9.3), trailing a column of white smoke, the spectators cheer as the noise of the launch shakes the ground.

Inside Orion, the crew is buffeted by the violence of the launch and deafened by the roar of the SRB, the cabin filled with yellow-white light from the rocket glare almost thirty meters below. Just one minute after launch, Ares I is travelling at Mach 1.5 at an altitude of more than ten kilometers, accelerated by the thrust of the SRB subjecting the crew to almost three Gs.

Once safely in low Earth orbit (LEO), the crew prepares for the rendezvous and docking with the TransHab that will provide living quarters during the journey and the In-Space Propulsion Stage (ISPS) that will provide them with the boost, sending them on their way to Mars.

TRANS-MARS INJECTION AND INTERPLANETARY TRAVEL

During the first few days of the mission, the astronauts spend time becoming acquainted with their interplanetary home away from home, trying not to collide with equipment or dislodge switches, and making sure to use the foot restraints and Velcro when preparing meals or performing experiments.

Three or four days following trans-Mars injection (TMI), when astronauts peer out of the triple-paned observation window, the Earth is reduced to a blue bowl of light, although it still shines bright enough to read a book by. At this distance, the vehicle also begins to pass out of the Earth's sphere of influence, an imaginary point in space where the gravitational potential of Earth is no longer the dominant force.

One of the daily rituals the pilot performs is checking the trajectory of the vehicle as it heads for Mars. If the flight path is found to be misaligned, it will be necessary to make a trajectory correction maneuver (TCM) based on corrective burn parameters calculated by mission control and also information provided by the inertial navigation system (INS) onboard the vehicle.

With the vehicle safely en-route, the crew settle down to daily life onboard the TransHab. It was decided the best approach to illustrating the day-to-day routine operations of the crew and the factors affecting life onboard an interplanetary spaceship was to describe these activities item by item.

Life and death

Given the seemingly endless list of possible contingencies, it is inevitable the crew will become familiar dealing with emergencies. Some off-nominal events will be benign, others false alarms and most will be non life-threatening but some may have the potential to threaten the crew and the mission.

Perhaps the most worrying of all the threats is the one posed by a solar proton event (SPE), a burst of potentially lethal radiation that will trigger the automatic flare alarm. Although Orion and the TransHab are designed to shield the crew from

Figure 9.3 Ares I launch. (NASA.) See Plate 12 in color section.

solar and galactic radiation, the intense doses resulting from a SPE may be a life-threatening event. During the SPE astronauts will retreat to the storm shelter onboard the MTV and spend much of their time monitoring their personal dosimeters closely, to gauge how much radiation their bodies are absorbing. If the storm is particularly violent, one of more of the crewmembers may suffer acute radiation sickness (ARS), resulting in nausea, pain and hemorrhaging from the mucous membranes of the nose. If the storm is prolonged, these symptoms could escalate and, ultimately, crewmembers could die.

In the event of a death onboard Orion, it will be the commander's responsibility to ensure an autopsy is performed, which will require standard X-ray documentation, and tissue samples for toxicological, bacteriological and biochemical analysis. The crew medical officer (CMO), whose task it will be to perform the autopsy, will also need to take samples from the deceased crewmember's liver, kidney, lungs, brain, hair, skin and cerebrospinal fluid. Needless to say, all of these tasks will be conducted in microgravity. At the time of writing, there are no radiation exposure standards for human exploration missions and if this is still the case twenty years from now, a commander faced with the death of a crewmember may receive instructions from mission control to preserve the body for legal evidence. How the body would be preserved, and how the crew would react to the body of a crewmember occupying the cabin for the remainder of the mission is difficult to say, but it would surely have a destabilizing effect. Even if the commander were instructed to eject the body, how would the crew respond? Would they hold a service and if so, what denomination? What would the procedures be for burial in space?

Sleeping
After a long day practicing abort procedures, studying Martian topography and performing inflight experiments, astronauts will look forward to a good night's sleep. As there is no up or down, crewmembers will likely either bunk in their living quarters or perhaps simply tether themselves to a bulkhead so they don't float around and bump into something. In common with the Space Shuttle and International Space Station (ISS) missions, astronauts will be scheduled for eight hours sleep, but with the excitement of being en-route to Mars, the dangers of radiation exposure and the close confines of the vehicle, it may be difficult to achieve this amount of rest.

Given the success and effectiveness of the mission will depend on the ability of astronauts to maintain vigilance and a high level of cognitive performance, the issue of sleep deprivation is a worrying one for mission planners. The human body was designed to sleep at night and be active during the day, a cycle governed by environmental cues in a process known as *entrainment*. However, in the spacecraft environment there is an absence of these environmental cues, which means disruptive signals are sent to the astronaut's brain, resulting in an impairment of the amount and quality of sleep. After several decades of research, it has been established that insomnia is a prevalent symptom among astronauts but, despite reams of data, very little is known about the effects of sleep deprivation during long duration missions.

Fortunately, there is an abundance of terrestrial sleep data gleaned from

Antarctic research stations and polar exploration which may help astronauts cope with the problems of interplanetary insomnia. In fact, based on the experiences of polar explorers, the greatest cause for concern may not be lack of sleep, but excessive sleeping, known as hypersomnia. For example, during the long Earth to Mars transit there will be little for the crew to occupy themselves, a condition that may result in some crewmembers withdrawing from interpersonal contact as a means of coping with the inactivity. In fact, in some cases, such as a contingency event postponing the landing on Mars due to a dust storm or an equipment malfunction, mission control may actually encourage excessive sleeping.

Similar events have occurred in the pursuit of polar exploration. For example, during the *Fram* expedition in 1893, Fridtjof Nansen and one of his crewmembers, Hjalmar Johansen, made a dash for the North Pole. After spending four months using dogsleds and kayaks, the pair came ashore on Franz Josef Land, a bleak and desolate island located in the northern Arctic Ocean. After building a hut from stones and walrus hides, they spent nine months isolated from human contact, their world illuminated by a blubber lamp and their activity restricted to conversation, the occasional hunt for food and sleeping. Yet despite the extraordinary circumstances of their confinement and isolation, Nansen and Johansen survived none the worse for wear, their experience serving as an encouraging example to Mars-bound astronauts who, while not planning to emulate the Norwegians, can take comfort in the fact that such austere conditions are at least survivable.

Although some individuals can sleep through the proverbial storm, most of us are affected to varying degrees by noise, and for astronauts trying to get some peace and quiet en-route to Mars, the problem of night time sounds will undoubtedly prove annoying. Not only will there be the sounds of visits to the waste management facility and the noises made by crewmembers inadvertently striking a bulkhead, there will also be the noises of the spacecraft's hull expanding and contracting as it spins like a barbecue spit to ensure equal heating of the vehicle.

Expedition clothes
The crew will probably spend most of their time wearing a fire-retardant, two-garment ensemble consisting of pants and shirts. How long they will wear the same set of clothes is open to question as there is no documented interplanetary mission protocol for clothing changes. Since there are no washing facilities onboard, crewmembers will bring several sets clothing with them, but due to space restrictions, there will obviously be a limit. For example, weekly changes of their mission uniform would total more than three hundred items and would impose a significant weight penalty, so it is likely crewmembers will only change twice per month. Although this may seem unhygienic, precedents have been set in polar exploration. Remember Nansen and Johansen? During the nine months they spent on Franz Josef Land the two Norwegians were unable to change clothes at all and instead had to attempt washing the garments in icy water every once in a while. No wonder they spent much of their time dreaming of warm baths and shops stocked with clean clothes!

Expedition astronauts will no doubt understand their participation in an enterprise as unique as a manned mission to Mars requires acceptance of hygiene that is perhaps less than optimal. Nevertheless, even with restrictions of clothing changes, it will be important each crewmember adheres to the highest standards possible since the subjective feelings of other crewmembers may easily be affected by the appearance of others who let standards slip.

Hygiene

"We have some special space shampoo that doesn't require water, and it does a pretty good job. So at the end of the mission, even though it's six months without a bath, we're still pretty good, and we don't smell too bad."

NASA astronaut Mike Fincke, explaining to Kingston Community School students in South Australia that keeping oneself clean in space is not that much of a problem, despite the lack of a shower or bathtub.

Hygiene standards are likely to contrast markedly with those deemed acceptable on current ISS and Space Shuttle missions. Although astronauts will still shave, trim their nails, brush their teeth and wash their face and hands regularly, the daily shower will either be a luxury activity or unavailable due to space restrictions. However, research conducted onboard nuclear submarines, the ISS and at Antarctic stations has demonstrated showers are important morale-boosters and it is likely spacecraft designers will be forced to provide the crew with some means of achieving an acceptable level of cleanliness. Given the cramped quarters of the crew vehicle it is likely a collapsible shower will be provided, requiring crewmembers to unstow and deploy the shower once or twice a week. Although unstowing, deploying and cleaning will take some time, the crew will probably consider it a pleasant diversion from the weekly routine, especially during the transit phases of the mission.

Running to Mars

Exercise has been recognized by generations of polar explorers for its therapeutic value in countering the depressing effects of isolation and confinement. Although the recreational value of exercise will no doubt be acknowledged by Mars-bound astronauts, the requirement to spend three or four hours running or cycling every day will also have other benefits for those confined to a microgravity environment. Evidence from dozens of long-duration space missions has confirmed astronauts subjected to long periods of microgravity lose not only bone mass but also muscle strength. The loss of bone mass is the greatest concern for mission planners since the bones of astronauts could become so weak they could fracture under Martian gravity. Astronauts assigned to long-duration increments onboard the ISS understand the insidious effects of bone demineralization and most adhere to a strict regime of exercise involving at least two hours of running on a treadmill or cycling on an ergometer. Even when following such a rigorous training regime, astronauts returning to Earth still lose as much as ten percent of their lower body bone mass. You would think losing so much bone mass would motivate crewmembers to exercise as much as possible but there is evidence from space missions that some crewmembers have not always followed the prescribed exercise program.

Although the physiological consequences to an ISS and Mars-bound astronaut who resist performing the required exercise may be similar, the medical consequences are not. An ISS astronaut who returns to Earth and suffers a fracture will have access to terrestrial medical facilities but an astronaut who suffers a fracture on Mars will only have access to limited medical equipment. Furthermore, the consequences of a crewmember being temporarily disabled will impose severe restrictions on the amount of work the crew will be able to perform during their surface stay.

Needless to say, if crewmembers are not motivated by the promise of avoiding increased susceptibility to fracture, mission planners will need to figure out other ways to encourage astronauts to exercise. One way of motivating the crew will be to make exercise recreational. To do this, some sort of mental diversion will be needed such as music, videos or implementing a competitive element in the form of an interactive training experience. For example, while cycling on the ergometer, astronauts could be presented with interactive 3D graphics of rolling green hills or painted-desert scenery, and real time representation of essential data such as power output, heart rate and current speed would be provided. For those astronauts with a competitive streak, videos of actual bike races could be downloaded to make the crewmember feel as if he/she was riding in the real event.

Due to the space restrictions onboard, it is likely only one or two items of exercise equipment will be available, one of which is most likely to be a bicycle ergometer because of the quality of workout it provides and the equipment's versatility. With three crewmembers, each requiring three to four hours of exercise per day, the exercise devices will be used intensively and will be subject to significant wear and tear. Such intensive use will demand at least one back-up device be provided.

Preparing meals

Exploration missions and prolonged periods of zero gravity present challenges for food preparation. The crew will not have a refrigerator or freezer onboard, so foods will be thermo-stabilized, irradiated, freeze-dried or canned. Due to the absence of friends, family and normal recreational pursuits, combined with the effects of prolonged isolation and confinement, food will assume added importance as crewmembers will focus on meals as a substitute for the customary sources of personal gratification on Earth.

Antarctic station managers and commanders of nuclear submarines recognize the importance of food in maintaining group morale and it will be important to cater to the preferences of each crewmember to avoid monotony. If the crew is an international one, the problems of stocking the crew vehicle's cupboard will be even more of a challenge but it is likely the crew will have access to about two hundred and fifty food and beverage items. Unlike Space Shuttle missions where preparation time is kept to a minimum, the food system designed for a Mars mission will not be designed for fast delivery as the crew will not be subject to the same time constraints and heavy workloads as Shuttle astronauts. In fact, the preparation of meals will probably be viewed as a pleasant activity and distraction from the otherwise mundane routine of exercise and inflight monitoring of data. Also, it is probable some astronauts will take advantage of the low-tempo operations and spend time preparing labor-intensive meals for their fellow crewmembers. In common with ISS operations and many polar expeditions, the crew will eat at least one meal together at a scheduled time, a practice proven to increase group cohesiveness.

Another practice that increased group cohesiveness during polar expeditions was the consumption of alcohol but its use has also been the bane of many Antarctic stations since the earliest days, due to heavy drinking. Whether alcohol will be provided to crews charged with safely executing the most dangerous and expensive mission in human history remains to be seen. Ultimately, the question will come down to human factors engineering, a specialization that considers the consequences of errors when designing precautions in a system such as a spacecraft. The human factors engineer will argue the more serious the possible consequences, the more safeguards must be implemented to protect the system. Although the astronauts might argue limited responsible use of alcohol would not result in significant problems, the human factors engineer would counter that a possibility exists for the effects of alcohol consumption to impair the response of crewmembers to an emergency requiring unimpaired cognitive abilities. Although the question of the availability of alcohol for Expedition Class missions will no doubt be a major topic of discussion among mission planners, astronauts and human factors engineers, it is likely alcohol will be prohibited during the first missions.

Working en-route

Perhaps the greatest hazard of many polar voyages was the boredom endured by the men during the long immobile periods. A similar problem may be faced by astronauts enduring the long interplanetary transits. In fact, mission planners will have to work hard to ensure boredom does not become a significant mission stressor. To do this, workloads will need to be carefully devised to ensure mission-related tasks are divided equally among crewmembers. Once again, the annals of polar exploration provide salient examples of how the burden of long-duration isolation can be offset by keeping crewmembers busy. Nansen, for example, kept his crew active throughout the long winter with housekeeping chores and vessel maintenance and it is likely mission planners will have a shopping list of activities for astronauts. Such tasks will probably include regular contingency simulations, activities preparing for landing on Mars, reviews of descent procedures and, of course, exercise. Yet, even when following a carefully scripted day-to-day list of assignments, astronauts will inevitably occasionally find themselves with nothing to do. Similar conditions occur at Antarctic research stations, where managers strive to provide workers with meaningful and non-repetitive tasks to alleviate boredom and its behavioral consequences. Usually, these tasks are goal-oriented, such as constructing a new building or studying for an exam. Although such a clearly defined set of goals is effective, the sheer length of time of confinement and isolation often results in some form of malaise affecting crewmembers at some stage of their increment. Prolonged periods with few work-related demands might have the potential to compromise the way a crew might respond to an emergency. Such an event would demand crewmembers make a rapid transition from relative inactivity to extremely high workloads, and it is possible the crew's response might be lethargic and even misdirected.

To avoid the aforementioned problems, mission planners will need to exert care in planning workloads and to ensure crewmembers receive extensive pre-mission training in the art of learning to be inactive without feeling guilty. Special attention will need to be paid to the slow phases of the mission when it will be necessary to keep the crew occupied with meaningful work.

External communication

Personal communication

During the mission, news will not travel as quickly as it did on previous LEO missions. Unlike all space missions to date, astronauts en-route to Mars will not have the luxury of two-way communication due to the lag time, which will increase as the mission progresses. In fact, by the time the astronauts arrive at Mars, the lag time may be as much as ten minutes. Due to the response lag, crewmembers will instead log personal communication messages from their private quarters. The messages will then be sent via a scrambler to ensure transmissions are kept private.

A problem that is perhaps greater than the communication lag is the issue of a crewmember receiving negative information from home, an event that could have potential catastrophic effect upon someone who is isolated and confined. However, astronauts on long-duration mission onboard the ISS have dealt with bad news with no apparent negative performance effects. For example, astronaut Dan Tani, received the news his ninety-year-old mother had died in a car crash while embarked upon a long-duration mission onboard the ISS. Since Tani had asked to be informed of any emergencies while he was away in space, Tani's wife and a flight surgeon on the ground broke the news in a video conference call. A few days later, Tani watched his mother's funeral via a video link while onboard the ISS, before deciding to continue with normal duties.

A similar event occurred in 1978, when cosmonaut Georgy Grechko, less than a month into a three month expedition onboard the Soviet Salyut 6 space station, received word his father had died. However, in Grechko's case, ground controllers decided to not to inform the cosmonaut until after he had landed and, rumor has it, the controllers even sent Grechko letters supposedly written by his father after his death, via supply vehicles. Clearly, such deception is unacceptable during any mission, and it will be important for mission planners and astronauts to agree upon policies regarding negative news.

Although a policy of not permitting crewmembers any personal communications is unlikely to ever be accepted by mission planners, it is worth noting one of the most successful expeditions in history received no news from home for more than three years. Once again, it was Nansen's Norwegian Polar Expedition which, despite maintaining communication silence for three years, experienced a minimum of disputes and performed its mission commendably.

Mission communications

In just about every expedition or mission in which a group of individuals is isolated and must communicate with mission managers, problems occur. Often, the problem concerns the group of isolated individuals complaining their distant mission managers have little comprehension of the constraints of living in such an isolated environment, a perception that has the potential to strain the relationship between mission control and the crew. Furthermore, since the isolated group represents a closed world with a unique frame of reference, as the mission progresses the group perceives itself as having less and less in common with ground control and so rifts inevitably appear, disputes accumulate and conflicts erupt. Examples of mission-related communications problems can be found in excerpts from a diary kept by a cosmonaut, Valentin Lebedev, during a seven-month flight onboard the Russian space station, Salyut 7.

> "The most difficult thing about this flight is keeping calm in dealing with ground control and with other crew members, because pent-up fatigue could generate serious friction."
>
> *Valentin Lebedev, Diary of a Cosmonaut: 211 Days in Space.[7]*

During their trip, there will be occasions when crewmembers will vent their interpersonal tension at mission managers as a means of displacing the frustration they will no doubt experience at some time during the mission. Such conflict between mission control and crew is a predictable and inevitable phenomenon observed in all isolated groups living in hostile environments, and certain measures can mitigate the effects. For example, the person tasked with talking with the crew will be an astronaut who will be able to establish confidence and a common bond with crewmembers. Also, the use of a two-way video capability and informal mission-related communications will facilitate communication etiquette and reduce the incidence of conflict.

Getting along

> "The conventional thinking is that on a long-term mission, you work your way through the first half, you get to the half-way point and then you say, 'but wait! I've got another half to go.' And then a temporary depression may set in."
>
> *NASA psychiatrist, Nick Kanas*

Some occupations are more stressful than others and the job of an interplanetary astronaut will probably be one of the most stressful of all. Fortunately, astronauts are trained to deal with stress better than almost anyone. Not only must they deal with the dangers of explosive decompression, slow death by radiation sickness, and equipment malfunctions, they must also deal with each other, living in close-quarters twenty-four hours a day for month after month. For a crew bound for Mars, the stress will be compounded by mission complexity, limited abort options and the sheer length of time away from home and family. Unsurprisingly, the effect of being cooped up with other members in what constitutes a tiny society for several months will exert additional stress upon crewmembers.

> "The curtain of blackness which has fallen over the outer world of icy desolation has descended upon the inner world of our souls. Around the tables, in the laboratory, and in the forecastle, men are sitting about sad and dejected,

lost in dreams of melancholy from which, now and then, one arouses with an empty attempt at enthusiasm."

Frederick Cook. 1898–99. Belgica Expedition, on the first winter spent below the Antarctic Circle.

While it is unlikely astronauts will suffer the same level of depression as the crew of the Belgica, it is inevitable that, as crewmembers become increasingly weary of one another, the potential for interpersonal conflict will increase. However, polar exploration has demonstrated the potential for such conflict can be minimized by eliminating the differences among the members of the expedition, as exemplified by Nansen's *Fram* and Shackleton's Transantarctic expeditions. For example, Nansen's crew endured three years of isolation and confinement, while Shackleton and his crew survived more than six hundred days in the Antarctic. Yet, despite the austere and demanding conditions imposed upon their crews, Nansen's and Shackleton's ventures were notable, not for any interpersonal conflict, but rather for the crewmembers' solidarity, as evidenced by this extract from Alfred Lansing's book, *Endurance* [6].

"It was remarkable that there were not more cases of friction among the men, especially after the Antarctic night set in. The gathering darkness and the unpredictable weather limited their activities to an ever-constricting area around the ship. There was little to occupy them, and they were in closer contact with one another than ever. But instead of getting on each other's nerves, the entire party seemed to become more close-knit."

Another reason the crews of Nansen and Shackleton performed so admirably was due to them living in an optimal organizational structure, governed by the situation-specific expertise of a clear and credible leader. As we shall see in the following section, special leadership and interpersonal skills are required to maintain authority and control under isolated and confined conditions.

Leadership

Although Ernest Shackleton (Figure 9.4) is universally recognized as one of the greatest leaders ever and one of the giants of Antarctic exploration, he was not actually part of any of the major successful explorations or discoveries in Antarctica. However, despite a lack of glorious "firsts", what Shackleton achieved during his expeditions stands, in the annals of exploration, as some of the greatest adventures ever recorded. His legacy of bravery in the face of adversity, of personal responsibility, sacrifice and exemplary leadership serve as a role model for all future Mars mission commanders.

On December 5, 1914, Shackleton and his twenty-seven crew embarked upon their mission to become the first to traverse the Antarctic continent. Almost immediately, their ship, the *Endurance*, was beset by walls of solid pack ice and repeated attempts over many months failed to free the ship. Eventually, the pressure of the ice resulted in the *Endurance* being crushed and the mission for Shackleton and his men became one of survival rather than exploration. As a fierce winter

Figure 9.4 The Anglo-Irish Antarctic explorer, Ernest Shackleton (1874–1922), best known for leading the Endurance expedition of 1914–16. (NOAA.)

approached, Shackleton mobilized his crew and planned their escape, which involved two attempts to reach open water by man-hauling three life-boats over the ice on crude sledges. On finally reaching open water, Shackleton and his men made a dangerous six-day journey to the temporary safety of a windswept and desolate island known as Elephant Island.

Shackleton realized rescue from Elephant Island was extremely unlikely, so he selected his five best men and set out on a treacherous 1300-kilometre journey across the Southern Ocean, considered by many mariners the most dangerous stretch of water on the planet. After enduring mountainous seas, and thanks to the extraordinary navigation skills of Frank Worsley, Shackleton reached South Georgia Island. After a trek across mountains that have only recently been charted, Shackleton reached Stromness whaling station after six hundred and three days away from civilization. From Stromness, Shackleton led the rescue effort to save his crew who had remained on Elephant Island. All twenty-seven crew who started the expedition survived the ordeal.

Given the significant risks faced by Mars-bound crews, the difference between success and failure and between living and dying may be determined by the strategies employed by the expedition leader and, in the event of a life-threatening event, it is likely most crew will hope for a Shackleton to guide them through adversity and uncertainty. But how did Shackleton address the unique demands of his crew and his situation?

First, Shackleton never lost sight of the ultimate goal of surviving and, to this end, he directed his energy and the energy of his crew towards activities required to stay alive. Secondly, Shackleton possessed an innate ability to inspire such levels of confidence and instill optimism in his men that they truly believed they could overcome any obstacle that came in their way. Thirdly, the power of a cohesive team was not lost on Shackleton, who was a master at devising ways each crewmember could contribute and feel a valued member of the team. Fourthly, although he was called "the Boss", Shackleton went out of his way to ensure he was given the same treatment as the rest of his crew and even went so far as scolding the cook who had prepared him a special meal! By minimizing his status, Shackleton ensured his expedition was a classless society, a fact exemplified by an extraordinary level of mutual respect among his crew. For example, when on the brink of starvation, a crewmember spilt his ration of milk, each of his tent mates poured some of their own rations into their companion's mug.

Despite the austere and pitiful conditions, Shackleton was also able to successfully encourage humorous and lighthearted activities, throwing a party for the crew halfway through the two-month period of constant darkness. 'The Boss' also knew when to take big risks. For example, during their traverse across South Georgia Island, Shackleton and the two accompanying crewmembers found themselves stranded on a glacier at almost fifteen hundred meters altitude. With night falling and fog rolling in, to remain on the glacier would have meant a slow death by hypothermia but the only alternative was a risky slide down the glacier in darkness. Creating a makeshift toboggan out of their rope, Shackleton and his team slid a harrowing six hundred meters into the darkness, arriving safely at the bottom of the glacier. Finally, and perhaps most importantly, Shackleton and his crew never gave up. Despite the isolation, solid pack ice, confinement, attacks by leopard seals and the ever-present threat of starvation, solutions were found and adversity was overcome.

Leisure time

An insight into the type of leisure activities astronauts might pursue is revealed in research conducted at Antarctic research stations, onboard nuclear submarines, in accounts of polar explorers and, of course, from previous long-duration space missions.

Based on these reports, one of the most popular off-duty activities is simply talking, with a tendency for reading to occupy more time as a mission progresses [2, 3]. Also high up on the list is watching movies, which happens to be the overwhelming favorite leisure activity onboard nuclear submarines. Given the role of recreation in maintaining psychological homeostasis during periods of stress, it is highly likely the crew will have access to extensive libraries that will include literature, prerecorded programming, feature films, educational materials and special broadcasts.

One recreational activity that may be implemented en-route to Mars is the weekly lecture series, a tradition established by polar explorers. The subjects of these lectures could include topics of interest such as Martian geography, the finer points of aerocapture and instruction in assisting the crew medical officer in the event of a crewmember suffering a rapid decompression event.

Habitability

Every Mars mission design study conducted has evaluated the need to provide crewmembers with personal space and the results of nearly every design have included the provision of private sleeping quarters for each crewmember. Unfortunately, space onboard spacecraft has always been at a premium and this problem is unlikely to be resolved in the near future. However, despite extensive research conducted on winter-over personnel, polar explorers and other humans routinely subjected to confinement, there is no agreement on what is considered adequate or optimal space. Fortunately, there are dozens of examples of personnel performing admirably under austere living conditions. Shackleton's men survived for months in extremely crowded conditions, and Nansen and Johansen endured each other while cooped up in a tiny hut for nine months.

In-flight medical care

A manned Mars mission will be one of great medical significance and it is probable at least one of the crew will be a physician with surgical training, and the other crewmembers will have received extensive CMO training. This training will be necessary since the weight and volume restrictions on the vehicle will severely limit the availability of surgical and anesthetic equipment to cover all but the most likely situations.

Historically, illness and injury have accounted for more failures and delays to missions than defective transportation systems and, during a Mars mission, the potential for a medical contingency to impact the mission will be significant for a number of reasons. First, there is the long communication delay, which may last anything up to forty-six minutes, depending on the orbital configuration of the Earth and Mars. Second, there are the limited medical care resources and the limited crew training and experience to consider and, thirdly, there is the abnormal pathophysiology of spaceflight, making any surgical intervention fraught with risk.

Medical care will also extend to ensuring the behavioral health of crewmembers, a requirement involving a team of NASA psychiatrists and psychologists with extensive experience working with astronauts and their families during missions. Much of the monitoring of astronauts' behavioral health will be by means of private psychological conferences (PPC) and CMO-administered psychological diagnostic tests. The results of the diagnostic tests will be sent to the crew surgeon at mission control for review and any recommended therapeutic response. Some might question the need to monitor the crew so closely, arguing that the crew selected to travel to Mars will most likely be the most thoroughly scrutinized and rigorously screened humans in history. However, while a serious or mission-threatening psychological incident will be unlikely, the consequences of a severe psychosis would be catastrophic on an expedition with no abort capability.

This section has described the features of life onboard a spacecraft while travelling to and from Mars. The challenges of living in such an environment for so long may seem demanding, but no more so than the hardships endured by hundreds of crewmembers of polar expeditions more than a century ago. The voyage will be long and arduous but the worry generated by the ever-present risks will surely be offset by the sheer scale of the adventure and the adrenaline rush that will come from waking up every morning with the realization that, in the best traditions of exploration, this is something no human has done before. As the crew approaches Mars, only the hazard of entry, descent and landing (EDL) remains but, of all the expedition dangers, this is potentially the most deadly.

ARRIVAL AT MARS AND ORBIT CAPTURE

Mars Orbit Insertion (MOI) is a procedure that must be performed to place the vehicle into a stable orbit around Mars and is a one-shot opportunity that, if missed, sends the vehicle either crashing into the surface or flying past Mars into deep space. To ensure safe capture into orbit, the MOI procedure requires a series of activities planned weeks in advance culminating in a dense sequence of mission critical activities, commencing nine days prior to arrival with a fine targeting maneuver. Fortunately, for the crew, the vehicle is designed with a high degree of autonomy, enabling the spacecraft to resolve most MOI operations and contingencies without crew intervention.

To conduct the necessary fine targeting maneuvers and MOI burn, the vehicle uses its Reaction Control System (RCS) thrusters. The burn is only initiated after the onboard computers check information provided by the star trackers, Doppler system and a delta differential one-way ranging (DDOR) system, confirming the vehicle's position relative to Mars by a procedure called triangulation. Once the vehicle's position is confirmed, the attitude and orbit control system (ACOS) is updated by the computer and a trajectory is calculated to place the vehicle in a pre-defined narrow entry corridor for Mars orbit. Once the final retargeting maneuver is performed, the computer programs the capture maneuver and configures the vehicle for capture configuration, which involves updating the onboard Failure Detection Isolation and Recovery (FDIR) settings and directing the RCS to provide optimal thrust direction. Meanwhile, the Commander and Pilot review possible failure scenarios and review recoverable and non-recoverable outcomes that may occur depending on engine performance or trajectory alignment. For example, the crew must be prepared for an off-nominal situation such as a quasi-insertion event in which the vehicle finds itself in too large an orbit requiring execution of additional maneuvers.

MARS ENTRY, DESCENT AND LANDING

The United States has successfully landed six robotic systems on the Martian surface. Each system had a landed mass of less than six hundred kilograms, whereas most manned mission architectures require landing at least forty tonnes! These facts will definitely be on the minds of astronauts as they approach Mars and prepare to conduct the challenging entry, descent and landing (EDL) procedures. The EDL sequence of events, described in detail in Chapter 7, will include an approach phase, an entry/atmospheric deceleration phase, a parachute descent phase, a power descent phase and, finally, touchdown.

Approach phase

Although the MDH is transporting a crew of three, only two are active in the control loop. These crewmembers are responsible for supervising the automation of events in the EDL phases, although the astronauts still retain the option of reverting to manual control in the final stages of the descent phase and also to re-designate the landing site. In all EDL phases prior to powered descent, switching to manual is considered an off-nominal procedure and is only performed if the automation is not working properly or if some unexpected event occurs requiring human intervention. Such a level of automation is a design requirement based on the technical experience and confidence gained from uninhabited aerial vehicle (UAV) technologies enabling the human operator more time to focus on tasks such as dealing with unusual situations requiring judgment and reasoning skills.

The first stage of EDL is to slow the vehicle so it can be placed into what mission planners refer to as a *Mars staging orbit*. One way they may do this is to perform an aerocapture maneuver, a part of a family of aero assist technologies enabling the vehicle to decelerate into a Mars orbit either by rocket propulsion, which requires propellant, or by atmospheric friction. The difficulty of entering Mars orbit from an interplanetary trajectory while travelling at several kilometers per second requires precise navigation by the pilot and subjects the vehicle to intense heating.

To start the aerocapture maneuver, the Pilot selects a feasible aerocapture trajectory from the navigation computer ensuring the vehicle follows the optimum flight path angle and flight corridor. The trajectory is one that does not subject the deconditioned crew to more than five G and results in a sufficient deceleration to slow the vehicle into Mars orbit. The angle is important because, if the angle is too steep, the vehicle will be subjected to high deceleration loads and intense heating. In fact, if the angle is much too steep, the vehicle may burn up before even reaching the surface. Alternatively, if the angle is too shallow, the vehicle will not decelerate sufficiently and will not enter orbit.

Entry/atmospheric deceleration phase

Once the vehicle begins to encounter the Martian atmosphere, its guidance logic is automatically activated. The guidance system calculates the bank angle required to steer the vehicle's lift vector so, when the time comes for deploying the parachute, the vehicle is in exactly the correct position relative to the intended landing site. While the guidance system calculates the bank angle, the Pilot monitors the velocity of the vehicle in preparation for the next major event: the parachute deployment and descent.

Parachute descent phase

The pilot monitors the velocity, waiting for the Mach indicator to reach Mach 2.2, while the other crewmembers brace themselves for what is the most violent phase of the flight. At Mach 2.2, the supersonic drogue parachute is deployed automatically by the guidance system, and the crew feels as if they've been punched in the back as almost five Gs are instantly transmitted through their bodies. For the crew, deconditioned by months of exposure to zero gravity, the five Gs feel more like ten, the parachute deployment serving as a vicious reminder of just how much their bodies have deteriorated.

Gradually, the parachute decelerates the vehicle to subsonic speeds, allowing the crew to breathe again. The Pilot, eyes glued to the MFDs, pays particular attention to the Mach meter, waiting for it to scroll down to Mach 0.8. At this velocity, the backshell and supersonic parachute are jettisoned and a larger, thirty-meter-diameter subsonic parachute is deployed.

The subsonic parachute is designed to reduce the velocity of the vehicle to forty of fifty meters per second. While the subsonic parachute does its job, the Pilot initiates terrain-relative navigation, using the landing radar to determine the vehicle's surface-relative altitude and velocity. The landing radar activates at an altitude of about 3700m and informs the Pilot if the vehicle is on course. Once the vehicle has descended to between 1500m and 1000m, the scanning light detection and ranging (lidar) radar begin to generate elevation maps of the Martian surface in the vicinity of the guidance system's planned landing site.

As lidar elevation maps are generated onto the MFDs, the Pilot checks to see if potential hazards exist near the landing site and runs through his/her contingency checklists, in case it becomes necessary to plan for an alternate location. To help him/her, the navigation system uses hazard identification logic designed to scan the area for any possible dangers.

Powered descent phase

Once the Pilot receives confirmation from his/her displays and the navigation system that the landing site is in fact safe, he/she commands the vehicle's guidance system to separate from the subsonic parachute and begin the powered descent.

During the descent phase, the crew is responsible for landing control, situational awareness (SA) and system status monitoring. Landing control is the responsibility of the Commander who ensures the automated systems are performing their assigned tasks and the vehicle is following the designated trajectory at the correct velocities [1]. In the event the automated systems are not working, the Commander switches to manual control to complete the landing phase. While the Commander is attending to his tasks, the Pilot monitors subsystems using the system status display to check for any anomalies, a procedure he/she conducts using checklists. Meanwhile, the remaining crewmembers remain vigilant throughout the descent, using their SA to check for any surface hazards the Commander or Pilot may have missed.

During the final phase of descent, the task-loading on the Commander and Pilot becomes progressively more demanding. The Commander checks intended trajectory against previous and predicted trajectory, monitors time to touchdown, checks for hazards and obstacles, checks terrain he/she can see with the terrain database, keeps an eye on fuel consumption, processes feedback from the Pilot, and prepares to notify the Pilot of any expected actions he/she may be required to take. Some of this information is presented on displays, some appears on monitors showing feed via externally mounted cameras, while still more data is presented in synthetic vision displays (SVDs). In the event of dust obscuring the cameras and windows, the crew can still land using the SVD.

Throughout the entire descent phase, the Commander and Pilot are guided by flight rules ((Table 9.4) and descent and landing failure modes (Table 9.5).

Table 9.4 Flight rules for landing on Mars[1]

General and Authority Rules for Landing

- The Mars Descent Vehicle must retain redundant capability in critical systems throughout the landing sequence, otherwise the mission must be aborted.
- The Commander has final authority for Go/No Go for initiating any burn.
- The Commander must make the final decision to abort.
- The Commander has final authority over touchdown site redesignation.
- The Commander may take over manual control at any time.
- If the Commander is incapacitated for any reason, the Pilot shall assume his place.

Mission Segment Rules: DOI

- Landing gear must be fully extended and locked prior to DOI.
- DOI will be aborted for any of the following:
  - Attitude deviations greater than nominal parameters as directed by GN&C
  - Rates greater than nominal parameters.
  - Overburn in excess of nominal parameters.

Mission Segment Rules: Descent Orbit Coast
- Any residual rates must be nulled.
- MDV orbit will be confirmed with Mission Control.
- MDV checkout must be completed ten minutes prior to PDI
  - Additional orbits are acceptable to comply with this rule.

Mission Segment Rules: PDI

- PDI will be initiated automatically to assure accurate thrust vector alignment and spacecraft attitude.
- Powered Descent will be aborted for any of the following:
  - Attitude deviations exceeding nominal parameters.
  - Rates greater than nominal parameters.
  - Uncorrected deviations outside the trajectory boundary.

Mission Segment Rules: Hover/Touchdown

- Voice communications between Commander and Pilot have top priority.
- An ascent stage abort to orbit will be performed for any of the following:
  - Rates greater than nominal parameters.
  - Vertical velocity greater than nominal parameters.
  - Horizontal velocity greater than nominal parameters.

Table 9.4 *cont.*

Mission Segment Rules: Post-Touchdown

- Completion of safing procedures has top priority.
- Ascent stage abort to orbit for any of the following:
  - Failure to shut down and safe descent engines.
  - Failure to safe any other part of the MDV.

[1] Adapted from Apollo Gray Lunar Landing Design

Table 9.5 Descent and landing failure action rules[1]

Malfunction	Flight Phase	Action
Loss of 1 Descent Engine	DOI	Continue. Attempt to restart.
	Powered Descent Burn 1	Continue. Attempt to restart.
	Powered Descent Burn 2	Continue. Attempt to restart. Abort to landing
	Hover	Continue. Abort to landing.
Loss of 2 Descent Engines	DOI	Continue. Attempt to restart. Must be resolved prior to Powered Descent Initiation (PDI).
	Powered Descent Burn 1	Attempt to restart. If unsuccessful, abort to orbit.
	Powered Descent Burn 2	Continue. Attempt to restart. Abort to landing.
	Hover	Continue. Abort to landing.
Descent Stage Fire	Any	Ascent stage abort to orbit
IMU Failure	DOI	Continue. Must be resolved prior to PDI.
	Powered Descent Burn 1	Continue/abort to orbit.
	Powered Descent Burn 2	Continue. Use landing radar if possible or abort to orbit.
	Hover	Continue or abort to landing under manual control if necessary.
Probable Loss of Ascent Engine	DOI	Continue. Troubleshoot during descent orbit coast.
	Powered Descent Burn 1	Descent stage aborts to orbit.
	Powered Descent Burn 2	Descent stage abort to orbit until fuel too low for descent stage abort, then abort to landing.
	Hover	Abort to landing.

Table 9.5 *cont.*

Command Software	DOI	Continue to descent orbit and trouble shoot if possible, otherwise abort to orbit.
	Powered Descent Burn 1	Troubleshoot or abort to orbit.
	Powered Descent Burn 2	Abort to landing or abort to orbit.
	Hover	Abort to landing.
Power System Failure	DOI	Switch to backup. Continue descent to orbit. Troubleshoot before PDI.
	Powered Descent Burn 1	Switch to backup and continue.
	Powered Descent Burn 2	Switch to backup and continue.
	Hover	Switch to backup and abort to landing.

[1] Adapted from Apollo Gray Lunar Landing Design

As the descent continues, the crew listens intently to the altitude being called out and confirmation from the Commander and Pilot the descent is still 'Go' for landing. As they descend through the Martian atmosphere, the abort options become fewer and their commitment to landing greater. At six thousand meters altitude, the crew hears the annunciation *Radar Lock,* informing them the vehicle is receiving information from the ground-based radar. Next is the Pilot's announcement of *High Gate* [8]. At this point the Commander sees the landing site for the first time. To confirm the vehicle is indeed coming up on the correct landing site, the Commander checks photographs taken by orbiting satellites. If the landscape looks unfamiliar, the Commander and Pilot run through whatever abort procedures are still available, knowing that by the time the vehicle is three thousand meters above the surface, the fuel remaining in the engines will be barely sufficient to abort to orbit. As the lander continues to descend, the computer revises the trajectory continuously as information about wind speed, height and velocity is received from the radar. To ensure the vehicle remains on the correct trajectory, the thrusters and attitude jets work overtime, pulsing short bursts every second. At one hundred and fifty meters, the Pilot calls *Low Gate*, indicating to the Commander the point where an assessment has to be made to select either automatic or manual control [8]. The Commander elects to fly the remainder of the descent under manual control. As the vehicle approaches the surface, visibility is obscured by Martian dust, forcing the Commander to rely on SVDs and radar. Finally, just two meters above the surface, the Commander cuts back the engine power, to prevent the back pressure from its own exhaust blowing itself up, and the vehicle falls gently to the surface.

INITIAL SURFACE OPERATIONS

Immediately after landing, the Commander runs through the post-touchdown checklist and a decision to stay/no stay is made based upon the integrity of the landing site and any damage to the vehicle or its systems. The crew then prepare for a period of adaptation to Martian gravity and their first steps on the Martian surface. To bodies adapted to life in zero gravity, becoming accustomed to walking around the vehicle is disorienting, but after three or four days the crew prepare for the first Martian EVA. The most articulate member of the crew egresses first and, holding tightly onto the handrails of the ladder, makes his/her way slowly to the surface. After months confined by the tight, curving walls of the vehicle, the sense of scale is overwhelming but through the one hundred and eighty degree vision of the faceplate, the crewmember sees the end of the ladder and steps off the rung onto the surface of Mars.

LONG TERM SURFACE OPERATIONS

In keeping with spaceflight tradition, the day begins with a musical wake-up call from Mission Control. Every once in a while, a VIP is patched through to the astronauts to wish them well before the start of the working day. After the daily ritual of showering and eating a breakfast of reconstituted cereal and Starbucks coffee, the crew prepares for planned surface activities. After donning their spacesuits, two crewmembers enter the egress chamber and pass through the airlock, before emerging onto the surface of Mars while the third crewmember remains inside, preparing to support and coordinate surface activities (Table 9.6).

One of the first tasks is to unstow the drill rig and begin the search for water and evidence of life. By extracting core samples (Figure 9.5) from various sites over a period of several weeks, the crew characterize the local area, before moving farther afield using the unpressurized rover (Figure 9.6). After six to eight hours performing their surface activities, the astronauts head home to the habitat. As they enter the habitat, the crew conducts a well-rehearsed ritual of brushing excess dust from their space suits. Once the bulk of the dust has been removed by brushing, an air hose is used to remove smaller particles. Any remaining dust is removed by the negative pressure airlock and the magnets located around the threshold of the airlock. Back inside, the astronauts relax in the small but comfortable confines of their habitat, helping each other prepare the evening meal.

A full account of surface operations and exploration activities is described in Chapter 10.

DEPARTURE PREPARATIONS AND DEPARTURE

Just like a holiday on Hawaii or Tahiti, the time to go home arrives much too soon. During their final days on Mars, the crew conducts predeparture operations including cleaning up the habitat, disposing of trash, placing systems in standby or off mode, and verifying ascent vehicle systems. On the final day on Mars, the crew uses its final EVA to load samples into the cargo bay, and then prepares for return to Martian orbit. Acutely aware the launch represents the first time humans have launched from Mars, the crew nervously waits for the comforting roar of the engines. As the ascent stage lifts the crew to orbit, they begin preparations for the rendezvous and docking with the TransHab.

Table 9.6 Scientific and human interest-driven activities

Scientific Activities

Activity	Indoor	Outdoor	Robotic
Geology	Rock analysis Geochemistry Sample storage Age dating Teleoperate rovers	Field Geology Mapping Geomorphology Stratigraphy Drilling	Sample collection Aerial reconnaissance Local resolution maps Multispectral mapping
Geophysics	Displays Data Analysis System Operations	Active seismic Electromagnetic sounding.	Local regional geophysical network (seismic).
Climate	Evolved gas analyzer.	Hydrologic history Recent cyclic changes	Aerial reconnaissance
Meteorology	Display Atmosphere composition	Outpost meteorology station Tethered balloon	Regional network
Exobiology	Culture samples Planetary quarantine Back-contamination controls.	Explore promising environments Hydrothermal areas Deep subsurface Drilling	Robotic field work.

Human-interest-driven activities

Arrival on Mars	Crew connects power, assembles habitat and raises flag(s).
Outpost setup	Crew assembles structures, power systems, deploy radiators.
Health maintenance	Crew under routine medical surveillance requiring tests for radiation exposure, exercise capacity and bone loss.
Crew meetings	Daily discussions concerning recent and future exploration activities. Changes to exploration plan discussed.
Crew recreation	Sightseeing excursions on foot and by rover.

RENDEZVOUS, DOCKING AND TRANSFER TO EARTH RETURN VEHICLE

After the clinical surroundings of their surface habitat, the mold and fungi-covered interior of the TransHab is a little alarming but the priority is to perform the trans-Earth injection (TEI) maneuver and begin the journey home. There will be plenty of time for household duties on the way back.

Figure 9.5 Astronauts drilling for samples. (John Frassanito and Associates/NASA.) See Plate 13 in color section.

Figure 9.6 Astronauts conducting exploration using an unpressurized rover. (John Frassanito and Associates/NASA.)

TRANS-EARTH INJECTION AND INTERPLANETARY TRAVEL

Following a successful burn, the crew turns its attention to scrubbing the bulkheads of the TransHab, a task that occupies the first four weeks of the journey home. Three months later, after entering LEO, the deconditioned crew waves goodbye to the TransHab and enters the Orion capsule for the de-orbit burn and re-entry to Earth.

EARTH ENTRY, DESCENT AND LANDING

Following a traditional splashdown off the California coast, the crew is transported to the quarantine facility at JSC, before beginning the long and arduous road to recovery. Before the rehabilitation process begins, the crew is transported to the NASA crew quarters where they take their first real shower in more than four months and undergo preliminary medical testing. This is followed by a meal, a press conference and some much-needed sleep. Although the astronauts are exhausted, it's likely they will have difficulty sleeping during the first few days of their re-adaptation to Earth because it takes a while for their space-adapted minds to become accustomed to the one gravity environment. During their transit from Mars to Earth, the astronauts floated day and night and to move anywhere in the cabin, required just a push off a bulkhead using a force no greater than a gentle push of a finger. In contrast, back on Earth, the simple act of rolling from one side of the bed to other requires enormous effort, and astronauts attempting the daunting challenge of visiting the washroom feel so discombobulated they end up having to crawl on their hands and knees.

POST LANDING RECOVERY

For astronauts returning from Mars, the mission does not end on landing. More than two years of exposure to zero and reduced gravity have taken their toll on the crewmembers' bodies. Their bones are softer, their muscles weaker and their reflexes slower, a reality only too evident to them as they are carried away on stretchers immediately after landing.

Short-duration flights are defined as less than thirty days, whereas long-duration flights are usually defined as six months, which is the standard increment duration for a stay on the ISS. A three-year interplanetary trip to Mars, however, requires creation of a new category of ultra-long duration flights. To date, the longest uninterrupted stay in space has been 437 days by cosmonaut, Valery Polyakov and the longest cumulative time in space was 747 days over three flights by cosmonaut Sergei Avdeyev.

The long duration of a return trip to Mars imposes severe physiological adaptations upon the crew which, on return to Earth, are manifested as impairments requiring extensive periods of cardiovascular, musculoskeletal and neurovestibular rehabilitation.

Bone demineralization

Perhaps one of the most damaging impairments is the insidious process of bone demineralization (see Chapter 8). Studies of the long-term effects of microgravity on the bones of ISS crewmembers revealed astronauts, on average, lost more than ten percent of their total hip bone mass during a six-month mission [5]. Although much of the lost mass bone mass is replaced a year following a long-duration flight, the bone structure and density takes much longer. During the course of a long transit to Mars and back, astronauts can expect to lose much more than ten percent of bone mass in their hip bone, the site of the most devastating osteoporotic fractures in those with reduced bone mass and density. In fact, the lengthy duration of a round-trip to Mars may result in such severe and prolonged disability that it is potentially beyond the point of safe return to Earth [4]. Due to the risks associated with such bone loss, returning Mars astronauts will be subject to extensive imaging of their skeletons for several years to help doctors characterize the recovery of parts of the hip bone and changes in strength of the other bones.

Muscle atrophy
Although the astronauts will have exercised for three or more hours per day during the trip back to Earth, this will not have been enough to preserve their muscles. Particularly affected will be the antigravity muscles such as the ankle plantarflexors/dorsiflexors, knee extensors/flexors and lower-back muscles, all of which will be significantly atrophied [9].

Neurovestibular problems
After such a lengthy adaptation to zero gravity and reduced gravity, return to a gravitational environment causes head movement and vehicle acceleration dependent disorientation, in addition to vertigo and blurring of vision. Some astronauts have referred to this state of disorientation as feeling discombobulated, and it is certain those returning from Mars will experience similar sensory confusion, requiring some form of rehabilitation.

For at least four months after landing, crewmembers will perform extensive rehabilitation exercises at JSC, where the Astronaut Strength Conditioning and Rehabilitation (ASCR) specialists will help muscle and bone-depleted astronauts regain their strength (Table 9.7).

After several months of rehabilitation, most symptoms will eventually resolve but, it is likely, the side effects of bone loss will last much longer and, in that sense, for the first crew to travel to Mars and the first humans to become citizens of two worlds, the journey may have no end. However, after years training for this iconic, once-in-a-lifetime expedition and having traveled further than any humans in history, any side effects will surely be judged by the crew to have been worth the risk by not only reinvigorating the spirit of exploration but also transforming humankind into a true interplanetary species.

The golden age of polar exploration was largely characterized by expeditions pursuing exceedingly recondite goals, essentially geographical abstractions surrounded by expanses of frozen sea of no apparent use to anyone. However, realizing a manned Mars mission will have a profound beneficial impact on the people of Earth by finally validating the possibility that Mars may someday be a home for humans. To demonstrate whether an outpost and eventual settlement can be

Table 9.7 Postflight rehabilitation plan

Description	The postflight rehabilitation plan is a 3-phase plan designed to protect the health and safety of astronauts returning from missions to Mars and to actively assist in the crewmembers' return to preflight health and fitness levels.		
Schedule	Duration	Schedule	Personnel
	Rehabilitation Phase I 120 min/day. Assisted walking. Hydrotherapy, proprioceptive neuromuscular facilitation (PNF) techniques, massage, and light manual resistance exercises.	0 to 7 days post-landing	Astronaut Strength Conditioning & Rehabilitation Staff (ASCR) Crewmembers, Crew Surgeon.
	Rehabilitation Phase II 120 to 150 min/day. Assisted walking. Hydrotherapy, Agility and coordination tasks, light cardiovascular exercise. Massage, PNF techniques, flexibility and strength exercise.	8 to 30 days post-landing	ASCR, Crewmembers and Crew Surgeon.
	Rehabilitation Phase III 150 to 180 min/day. Agility and coordination tasks. Cardiovascular exercise. PNF techniques, massage, hydrotherapy and strength exercises.	31 to 60 days post-landing	ASCR, Crewmembers and Crew Surgeon.
	Rehabilitation Phase IV 90 to 120 min/day. Cardiovascular and strength training. Massage. Fitness testing once per week.	61 to 120 days post-landing	ASCR, Crewmember and Crew Surgeon.
Special Requirements	Crewmembers will perform rehabilitation on duty days only (5d/wk). Medical status checks will be performed once per week. The ASCR and Exercise Physiology Laboratory will make recommendations to the crew surgeon regarding rehabilitation progress and exercise certification of crewmembers.		
Notes	During each rehabilitation phase, crewmembers will be assessed using fitness tests to evaluate isokinetic function, oxygen uptake, agility and coordination and flexibility.		

sustainable, it will be necessary for successive crews to conduct surface activities directed at proving a capability for self sufficiency. To facilitate such a strategy, mission planners must provide the crew with carefully considered surface activities and the surface systems the explorers will use. These challenges form the subject of the final chapters of this book.

REFERENCES

1. Brand, T.; Fuhrman, L.; Geller, D.; Hattis, P.; Paschall, S., and Tao, Y. GN&C Technology Needed to Achieve Pinpoint Landing Accuracy at Mars. Charles Stark Draper Laboratory Inc., Cambridge, MA. AIAA-2004-4748. AIAA/AAS Astrodynamic Specialist Conference and Exhibit, Providence, Rhode Island, Aug 16–19, 2004.

2. Doll, R.E., and Gunderson, E.K.E. Hobby Interest and Leisure Activity Behaviour among Station Members in Antarctica. San Diego, California. U.S. Navy Medical Neuropsychiatric Research Unit. Unit Report No. 69–34. 1969.

3. Eberhard, J.W. The Problem of Off-Duty Time in Long-Duration Space Missions. 3 vols. NASA CR 96721. McLean, Virginia.: Serendipity Associates. 1967.

4. Holick, M.F. Microgravity-induced bone loss – will it limit human space exploration? Lancet;355:1569–70, 2000.

5. Lang, T.; LeBlanc, A.; Evans, H.; Lu, Y.; Henant, H.; Yu, A. Cortical and trabecular bone mineral loss from the spine and hip in long-duration spaceflight. J. Bone Miner. Res. 19:1006–12, 2004.

6. Lansing. A. Endurance. Basic Books. 2nd Ed. p42. 1999.

7. Lebedev. V. Diary of a Cosmonaut: 211 Days in Space. Bantam. 1990.

8. Sostaric, R.R. Powered Descent Trajectory Guidance and Some Considerations for Human Lunar Landing. AAS Guidance and Control Conference, Breckenridge, Colorado, February 307, 2007.

9. Vandenburgh, H.; Chromiak, J.; Shansky, J.; Del Tatto, M.; Lemaire, J. Space travel directly induces skeletal muscle atrophy. FASEB J;13:1031–8, 1999.

10

Exploration activities and surface systems

There has been no shortage of suggestions of how to send humans to Mars but little has been written about what astronauts will actually do once they have arrived safely on the surface or about the systems that will support exploration activities. Since the outcome of surface activities will ultimately determine whether Mars has the potential to sustain permanent human settlement, and because several mission profiles feature crews remaining on the surface for up to five hundred days, the subject of surface operations and systems deserves particular attention.

SURFACE EXPLORATION

Exploration strategy
The first consideration driving the exploration strategy is the choice of landing site since, if the mission lands in a safe but scientifically bland region, there is a risk the crew will exhaust all areas of interest in the first few weeks. This consideration leads to the next issue, namely of mobility, which is likely to be limited during the first missions since most pressurized rovers will only have a range of a few hundred kilometers and consumables to support the crew for ten days. Many of the most interesting sites may be deemed too dangerous or inaccessible, so it is likely the first mission to Mars will need to weigh the tradeoffs between optimal location and scientific return.

Selecting a location to explore
In deciding where to land, mission planners will select a location most representative of Martian geophysical history and planetary evolution, a site offering the best opportunity to search for resources and the best place to search for life. Ideally, the site should be scenic too, so astronauts can enjoy what little recreation time they have. The problem is, even if these selection criteria were applied to Earth, it would be impossible to select an optimum site because there isn't any place on Earth truly representative of the whole planetary or evolutionary history, and the same applies to Mars.

There are different approaches to solving the issue of site selection, one involving first establishing a base on either Phobos or Deimos, where astronauts would stay while robotic landers scouted out locations on the Martian surface, returning samples once in a while for the crew to study. Eventually, an ideal landing site would be selected and the crew would land on Mars. However, the primary objective of traveling to Mars is to establish an outpost and eventual human settlement, so the "space station near Mars" solution is not an ideal one. Furthermore, it is a solution that ignores the probability future unmanned orbiters will carry improved imaging technology, allowing for easier site selection.

Another suggestion is to make the entire surface habitat into a rover by putting it on treads, enabling the crew to simply drive from site to site, but this solution suffers from the mass penalty incurred by configuring a mobile surface habitat. Yet another solution integrates robotic and human exploration, by deploying rovers two years ahead of the manned mission. The rovers would land at different sites on Mars and begin collecting samples. Shortly before the manned expedition arrives, the rovers would make their way to the site of the landing and, when the crew arrives, they would have samples from several different locations waiting for them. The crew would then analyze the samples and decide which to send back to Earth. Of course, this mission plan results in the humans becoming almost superfluous since the robots would be doing most of the exploration!

Surface exploration considerations
Surface exploration by robotic vehicles and human explorers will include a variety of activities, ranging from observing and analyzing surface and subsurface geology and collecting samples, to performing experiments designed to assess the ability of humans to survive on Mars. To assist astronauts in their exploration activities, the outpost will be equipped with all sorts of mineralogical, chemical and atmospheric testing facilities, in addition to a whole fleet of semi-autonomous and autonomous robotic vehicles. Armed with their exploration equipment, astronauts will conduct field investigations locally at first but, as experience accumulates, the scope of human exploration will grow from local to regional.

Mission schedule
In between conducting exploration assignments, astronauts will be required to spend time performing maintenance and housekeeping tasks such as tending to the life support system and ensuring Martian dust does not compromise the habitability of the outpost. A suggested daily schedule may follow a time allocation similar to the one described in Table 10.1.

Table 10.1 Mars surface mission time allocation

Daily Time Allocation

Activity	Time (h)	Activity	Time (h)
Sleep preparation, dress, undress	1	Meal preparation, clean-up	0.5
Sleep	8	Eating	1.0
Hygiene, cleaning.	1	Recreation	1
Exercise	1.5	Earth Communication	0.5
General planning, reporting, documentation	0.5	Group socialization, meetings, health monitoring, health care	1
System monitoring, inspection, calibration, maintenance, repair	1	Work	8

Mission Time Allocation

Task	Time (d)	Excursion Activities
Outpost site preparation, construction and verification	90	Charge Fuel Cells
Recuperation	7	Check Vehicles
Local excursions Analysis Regional excursions Analysis	100	Load Vehicle Plan Excursion Drive Vehicle Navigate
Recuperation	7	Don Suits
Distant excursions Analysis Distant excursions Analysis	100	Pre-breathe Egress Unload equipment Set up drill
Recuperation	7	Operate drill
Distant excursions Analysis Distant excursions Analysis	100	Collect Samples In Situ analysis Take photos Communicate Disassemble Equipment
Recuperation	7	Load Vehicle
Flare Retreat	15	Ingress Clean Suit
Distant excursions Analysis Distant excursions Analysis	100	Stow suit Inspect Vehicle Secure for night Sleep, eat, hygiene, etc.
System shutdown and preparation for departure	60	

Surface objectives
Surface operations will encompass a spectrum of activities ranging from basic human survival to space human factors and the search for life. The primary mission objectives are summarized in Table 10.2.

The surface objectives listed in Table 10.2 are by no means exhaustive and achieving them will require not only a carefully considered and orchestrated mission timeline but also highly reliable surface systems, the common elements of which are described here.

SURFACE SYSTEMS

The surface systems required to support extended science and exploration operations consist of six major systems: a power system, extravehicular activity (EVA) systems, a life support system (LSS), a surface habitat, in-situ resource utilization (ISRU) equipment and surface mobility systems. Since this chapter is based on the SpaceWorks Inc. (SEI) architecture, it is assumed each system is pre-deployed to Mars and determined to be functioning, before the crew departs Earth. Also, since the first SEI mission does not feature ISRU equipment, the systems discussed here are representative of a second or third mission.

Power generation and storage
One of the first crew tasks will be to deploy a surface power system to supply the outpost, lander, habitat, ISRU plant, mobility and construction equipment, and science experiments with electricity. The unique environment of Mars poses many challenges for power systems, due not only to the great distance from the Sun and the attenuation of the sunlight through the atmosphere but also due to the frequent dust-storms and the corrosive, carbon dioxide atmosphere.

Due to these difficult environmental conditions, Martian surface power is an obvious application for space fission systems, which are independent of seasonal and geographic sunlight issues such as solar arrays, which require the area of several football fields to produce the level of output required for even a small Martian base.

Concept of operations
A representative Fission Surface Power (FSP) system is characterized by four subsystems including a reactor, power conversion, heat rejection and power conditioning and distribution (PCAD). The system produces power by first transferring heat from the reactor to the power conversion and from the power conversion to the heat rejection. Electrical power generated by the power conversion is then processed through the PCAD to the user loads. The power for power conversion startup and auxiliary loads associated with the reactor and heat rejection is provided by the PCAD, which also provides the main communications link for command and health monitoring of the FSP system.

The FSP system is landed prior to arrival of the crew and located about one kilometer away from the habitat to reduce shielding mass. When the crew lands, a

Table 10.2 Surface mission objectives

Human Survival and Mission Duration	
Mission Objective	Rationale
1 Demonstrate human survivability on surface of Mars for 30 days.	Demonstrate mission architecture capability to support extra-terrestrial human habitation. Physical wellbeing of crew is primary mission requirement.
2 Minimize physiological effects of long-term exposure to Martian environment.	
3 Demonstrate human survivability on surface of Mars for 18 months.	Demonstrates mission architecture capability to support long-term extra-terrestrial human habitation.
4 Extend EVA operations to distances beyond 100km.	Regional and long-range EVA ensures greater science return.

Human Physiology and Psychology	
Mission Objective	Rationale
5 Characterize human interaction dynamics.	Provide greater understanding of space human factors in Martian environment.
6 Investigate and characterize human performance quality and reliability as a function of time.	Provide greater understanding of space human factors in Martian environment.

Life and Planetary Science	
Mission Objective	Rationale
7 Search for water within EVA radius of outpost.	Provides in-situ water source and provides insight into evolution of Mars.
8 Search for carbon within EVA radius of outpost.	Provides in-situ source of carbon and provides insight into biological, meteorological and geological evolution of Mars.
9 Characterize climate variations within EVA radius of outpost.	Enables prediction of weather patterns and provides insight into Martian climate evolution.
10 Investigate surface and sub-surface geological characteristics within EVA radius of outpost.	Identifies potentially useful in-situ compounds and provides insight into surface evolution.
11 Evaluate effectiveness of excavating and utilizing Martian regolith as radiation shield material.	Enables evolved surface architecture and reduced system mass.
12 Search for sub-surface volatiles, geological structures and geothermal activity.	Provides insight into interior structure, thermal properties and geologic age of Mars.
13 Locate and catalogue useful resources within EVA radius of outpost.	Identifies areas of scientific interest and resource-rich deposits for future utilization.
14 Conduct remote scouting missions utilizing autonomous rovers.	Identifies future sites of interest and demonstrates autonomous rover capability.

power transmission cable is used to connect the FSP system with the habitat, and an autonomously deployed, vertical radiator is extended via a truss structure. Alternatively, the crew emplaces the FSP system by burying the reactor and moving regolith to form a berm, serving as a shield from radiation.

Design
The FSP system comprises a stainless-steel, uranium-dioxide fuelled reactor cooled by sodium potassium and pumped water heat rejection. Power conversion is provided by a Stirling convertor. The deployed span of the two symmetric radiator wings is approximately thirty-four meters from end to end. The bottom edge of the radiators is one meter above the surface to reduce the accumulation of dust on the radiator surfaces. The reactor core consists of eighty-five uranium dioxide fuel pins, housed in a stainless steel vessel approximately twenty centimeters in diameter.

Extravehicular activity
The core of a manned Mars mission is exploration. To achieve exploration goals, astronauts will search for resources, learn how to work safely in a harsh environment, and explore the lunar surface, activities (Table 10.3) requiring sustained periods of EVA outside the protective environment of the outpost.

Table 10.3 EVA tasks during Mars surface operations

EVA	Description	EVA	Description
Site Preparation	Survey and stake-out. Rock removal & smoothing. Establish navigation aids.	Habitat Installation	Unload, locate, level. Deploy & inflate back filling.
Shielding Installation	Regolith bagging for radiation protection. Bag stacking & clearing access paths.	Science	Sample collection. Installation of experiments. Location of mapping for geological survey.
Power Systems & Thermal Control System	Site preparation. Unload equipment from lander & transport. Deploy & assemble. Radiator deployment & activation. Connect to distribution.	Logistics & Upkeep	Unpack & Transport. Transfer & Storage. Waste disposal. Storage of spent Recyclables. Inspection. Field checks & measurements. Replacement of systems. Repair of equipment.
Resource Operations	Resource process site set up (pressure vessel, plumbing, gas holding tanks, pumps, heat exchangers).	Lander Operations	Servicing/minor repairs. Refueling. Pre-launch & checkout. Relocation of fuelling depot.

Figure 10.1 Spacesuit engineer Dustin Gohmert simulates work in a mock crater of JSC's Lunar Yard wearing the Mark III spacesuit. (NASA.)

EVAs will require astronauts to wear a suit capable of protecting them from the Mars environment characterized by an extremely low atmospheric pressure, surface temperatures ranging from minus one hundred and forty to plus seventeen Celsius, radiation, and abrasive and magnetically charged dust.

In preparation for the return to the Moon in 2020, NASA is making improvements to the A7LB suits originally used for the Apollo EVAs, which were plagued by vision obscuration, dust contamination, thermal control problems, joint and seal failures, and mobility and flexibility limitations.

The current iteration is the Mark III Suit (Figure 10.1), also known as the H-Suit, a full gas pressure hybrid configuration composed of hard elements, such as a Hard Upper Torso (HUT), and soft components such as single-axis fabric elbows, knees and ankle joints. Additional areas of mobility include a rolling convolute waist joint and use of bearings in multi-axis joint systems. Entry is via a hatch on the back of the HUT.

It is possible the first astronauts on Mars will conduct EVAs wearing a full gas pressure suit derived from the Mark III, but such a suit has many disadvantages including awkward donning and doffing, pre-breathing and pressurization, limited flexibility and dexterity and a high metabolic requirement. More versatile and better suited to long EVAs is the BioSuit, a suit that may revolutionize human space exploration

Bio-Suit

Based on the concept of biomechanically and cybernetically augmenting human performance capacity, the Bio-Suit System (Figure 10.2), conceived by Professor Dava Newman of the Massachusetts Institute of Technology (MIT), is envisioned to function as a second skin by providing mechanical counter-pressure (MCP).

Whereas the lightest Mark III Suit weighs at least 40 kilograms, Newman's skin-tight, alternative Bio-Suit weighs as little as 10 kilograms! Constructed of spandex and nylon, the multi-layered suit hugs the body's contours like a second layer of skin, and its MCP technology ensures constant pressure is applied to the surface of the body. This pressure is needed not only to counteract the vacuum of the Martian surface and to maintain the body's homeostasis, but also to avoid blood pooling.

Maintaining an even pressure over the surface of the human body has, until now, been achieved by utilizing the bulky gas-pressurization systems embodied by the Mark III Suit. Thanks to new MCP technology that works along lines of non-extension, or those lines along the body undergoing little stretching as the body moves, a gas-pressurization system is no longer necessary.

Another advantage of the Bio-Suit over the Mark III Suit is suit compromise. If the Mark III Suit suffers a puncture, the crewmember must return to the outpost and undergo decompression, whereas a small puncture of the skin of the Bio-Suit requires only a bandage.

The Bio-Suit will also help astronauts stave off the debilitating effects of osteoporosis and general deconditioning, thanks to its ability to provide crewmembers with resistance levels to maintain muscle and bone integrity. Perhaps the most striking capability of the Bio-Suit is, by virtue its flexibility, its ability to enhance human performance on the Martian surface and thereby empower exploration.

Bio-Suit instrumentation

Martian EVAs will last six to eight hours and will involve crewmembers visiting isolated locations and performing increasingly complex activities, a combination of factors that has the potential to increase the distraction and fatigue of the astronaut. However, should an accident occur in an isolated location, or should a crewmember become excessively tired, having access to the information provided by a wearable sensor system might improve chances of survival. Fortunately, the Bio-Suit features a suite of such sensors.

Integrated into the Bio-Suit is a bioinstrumentation system modeled on the Operational Bioinstrumentation System (OBS) developed in the 1970s at the University of Denver for use during Space Shuttle missions. The original OBS consisted of a Signal Conditioner, an EVA Cable, a Sternal Harness, and three electrodes, a suite of components that will be significantly upgraded in the Bio-Suit, featuring biosensors integrated into the textiles. The biosensors measure biomedical

Figure 10.2 The revolutionary Bio-Suit. (Professor Dava Newman, MIT: Inventor, Science and Engineering; Guillermo Trotti, A.I.A., Trotti & Associates, Inc. (Cambridge, MA): Design; Dainese (Vicenza, Italy): Fabrication; Douglas Sonders: Photography.

signals such as heart function, oxygen consumption, and body temperature, while a suite of biochemical sensors provides information concerning body fluids, and dosimeters measure the local radiation environment.

Life support systems
One of the main reasons for sending astronauts to live on and explore Mars is to determine if humans are capable of not only surviving there but also of working productively. Obviously, to prove humans can indeed function effectively for extended periods on Mars, a robust and reliable LSS will be required.

Life support system challenges
One of the greatest challenges posed by extended duration missions on the Martian surface is ensuring basic life-support requirements can be met for long periods of time. The remote location of Mars combined with its challenging surface environment and reduced gravity level, presents unique challenges in terms of designing life-support methods. Compounding the challenge is the limited experience in the field of extended duration LSSs and subsystems which, to date, is restricted to submarines, the ISS, the Space Shuttle, and Earth-based biospheres.

Choice of system
Although it is uncertain which type of system will be used for the first expedition, mission planners will probably opt for a standard physical/chemical system typical of systems used on current spacecraft. Such a system relies on a combination of physical and chemical processes to remove impurities from the air and water. In subsequent missions a hybrid system may be chosen, combining the advantages of a bioregenerative system with the benefits of a cached stocks option. In the bioregenerative system, higher plant life species are used to provide food, revitalize air and purify food, while the cached stocks system makes use of ISRU equipment to produce air and water.

Regardless of which option is chosen, the LSS must provide the crew with a habitable environment, including clean air and water, solid waste processing, food processing, biomass production, thermal control, in addition to supporting interfaces.

Life support system functions
Before describing the Mars Surface Habitat ECLSS, it is useful to review the LSSs according to their function and their interaction with other systems and subsystems (Table 10.4).

The SpaceWorks Inc. (SEI) architecture is based on a crew of three conducting surface operations for five hundred days. To sustain the crew, the LSS must include air, biomass, food, thermal and waste elements and water, each of which interacts in a comprehensive overall system that maximizes recycling of waste products. Calculating the mass estimates for these components is difficult because the LSS for the Mars mission has yet to be designed so it isn't possible to factor in the recovery percentages of elements such as water and biomass.

Table 10.4 Life Support Subsystem Description [4]

Habitat Life Support Subsystems and Interfaces		
Subsystem	Description	Life Support Interfaces
Air	Stores and maintains the cabin atmospheric gases, pressure and composition and serves as a fire detection and suppression system.	Biomass, Food, Thermal, Waste, Water, EVA Support, Human Accommodations, ISRU, Integrated Control, Power.
Biomass	Produces stores and provides raw agricultural products to the Food Subsystem while regenerating air and water.	Air, Food. Thermal Control, Waste, Water, ISRU, Integrated Control, Power.
Food	Receives harvested agricultural products from the Biomass Subsystem, stabilizes them and stores raw and stabilized agricultural products, food ingredients and prepackaged food and beverage items. Transforms raw agricultural products into a ready-to-eat form via food processing and meal preparation operations.	Air, Biomass, Thermal, Waste, Water, EVA Support, Human Accommodations, Integrated Control, Power.
Thermal	Maintains cabin temperature and humidity within bounds and rejects collected waste heat to environment.	Air, Biomass, Food, Waste, Water, Human Accommodations, Integrated Control, Power.
Waste	Collects and conditions solid waste material from anywhere in habitat, including packaging, human wastes and inedible biomass.	Air, Biomass, Food Thermal, Water, EVA Support, Human Accommodations, Integrated Control, Power, Radiation Production.
Water	Collects wastewater from all sources, recovers and transports potable water, and stores and provides that water at purity for crew consumption and hygiene.	Air, Biomass, Food Thermal, EVA Support, Human Accommodations, ISRU, Integrated Control, Power, Radiation Production.
External Life Support Interfaces		
External Interface	Description	Life Support Interface
Extravehicular Activity Support	Provides life support consumables for EVA, including oxygen, water, food and carbon dioxide and waste removal.	Air, Food, Waste, Water, Human Accommodations, Integrated Control, Power.

Table 10.4 *cont.*

External Life Support Interfaces		
External Interface	Description	Life Support Interface
Human Accommodations	Includes crew cabin layout, crew clothing, laundering and crew interaction with life support system.	Air, Biomass, Food, Thermal, Waste, Water, EVA Support, Integrated Control, Power.
In-Situ Resource Utilization	Provides life support commodities, such as gases, water, and regolith from local materials for use in life support system.	Air, Biomass, Water, Integrated Control, Power.
Integrated Control	Provides appropriate control for all the life support system.	ALL
Power	Provides energy to support all equipment and functions within life support system.	ALL
Radiation Protection	Provides protection from environmental radiation.	Food, Waste, Water, ISRU, Power.

Another aspect difficult to predict is the LSS failure rate. One of the mistakes of many mission designs is the assumption high-performance LSSs will function flawlessly for the two to three year duration of the mission. However, experience onboard the International Space Station (ISS) has shown, despite extensive ground checkout, hardware failures still occur with frustrating regularity, so for a Mars mission it will be necessary to include a back-up cache of replacement units, effectively doubling the system mass!

Life support requirements
Bearing in mind these considerations, it is still useful to present best estimates of what may be required to support the crew during their five-hundred day stay on Mars. A detailed discussion of all the LSS elements is beyond the scope of this book, so the emphasis is directed at providing the reader with an estimate of the LSS masses based on recent research. First, we consider oxygen requirements (Table 10.5), which range from 0.78 kilograms per crewmember per day while resting, to 0.96 kilograms per day when conducting EVAs [2], a requirement equating to 1260 kilograms based on nominal activity for the three crewmembers.

The mass constraints of the Mars LSS problem is evident when tallying the estimates presented in Table 10.5 which reveals a forty-two tonne mass for just the water and oxygen requirement. However, these figures are raw data and do not consider the engineering systems, which require margins, spares and fail-safe performance, nor do they consider the equivalent system mass (ESM) that includes estimates of mass for power systems and cooling. When the ESM requirements for

Table 10.5 Fundamental life support requirements [2]

Oxygen Requirements (kg/Crewmember/Day)	
Category	Oxygen Requirements
Low Activity Metabolic Load	0.78
Nominal Activity Metabolic Load	0.84
High Activity Metabolic Load	0.96

Total Surface Requirement for 3 crew during 500d surface stay based on nominal activity metabolic load: **1260KG**

Water Requirements (kg/Crewmember/Day)	
Water	Requirement
Drinks	2.00
Shower (one per two days)	2.72
Urinal Flush	0.50
Oral Hygiene	0.37
Hand Wash	4.08
Laundry	12.47
Dish Wash	5.44
Food Processing and Preparation	TBD
Total Hygiene Consumption	25.58
Metabolic and related Consumption	2.0
Total Water Consumption	**27.58**

Total Surface Requirement for 3 crew during 500d surface stay: **41,370 KG**

oxygen, water, biomass, food, thermal requirements, waste, EVA support and human accommodations are considered, the estimated total LSS mass may approach one hundred tonnes for a crew of three [3].

Although the LSS requirements appear to pose a significant challenge to mission planners, NASA's Advanced Life Support (ALS) project is gradually advancing the technology of recycling water and air resources which, if successful, will drive down the largest single LSS mass. Additional savings may be achieved in the event of developing a capability to utilize indigenous water from Mars which may, in principle, eliminate the need to recycle water *and* oxygen on the surface.

Surface habitat

Inflatable structures
The SEI mission surface architecture utilizes a two-level rigid pressure vessel similar to the inflatable structures used for the In-Space Transfer Habitat/TransHab flight elements. In addition to being easily compressed and cheap to transport, inflatable

Figure 10.3 Bigelow Aerospace's Genesis II inflatable module in orbit. (Bigelow Aerospace.)

structures provide large living spaces and have already been demonstrated in the harsh environment of low Earth orbit (LEO). For example, in July 2006, Bigelow Aerospace launched Genesis I which, until the launch of Genesis II (Figure 10.3) the following year, represented the most successful use of an inflatable structure in space.

TransHab project

The inflatable design was initially developed by NASA's TransHab project which first proposed the inflatable module concept as a crew quarters for the ISS. Inside the Space Shuttle's cargo bay, the module would have had a diameter of 4.3 meters but, once inflated, it would have expanded to 8.2 meters, providing the crew with a 340-cubic-meter facility. The TransHab module originally designed for the ISS was to have been a unique hybrid structure, combining the mass efficiencies of an inflatable structure with the advantages of a load-bearing hard structure. Unfortunately, in

2001, due to the $4.8 billion ISS budget deficit, Congress restricted NASA from spending any money on TransHab beyond the design studies. After NASA abandoned the TransHab technology, Robert T. Bigelow's company, Bigelow Aerospace, decided to refine the concept.

Bigelow started by purchasing a building near Ellington Field in late 2004 and, with the help of employees of Johnson Space Center's (JSC's) Structural Engineering Division, transformed the building into an inflatable production facility. Bigelow also licensed the rights to the patents for several TransHab technologies and materials and arranged Interpersonal Act Agreements with JSC's inflatable module experts, Jason Raboin, Chris Johnson, Gary Spexarth and Glenn Miller, permitting them to work full-time with Bigelow Aerospace. The culmination of Bigelow's work was the flight of Genesis I and II, demonstrating the potential of multipurpose habitable space structures. NASA, having seen the success of Bigelow's efforts, is now encouraging the company in its development of the inflatable structures with a view to possibly using them as habitats on Mars.

Inflatable structures in Antarctica

Until Bigelow's inflatable structures are developed for Mars applications, NASA is pursuing other inflatable options. Recently, NASA, the National Science Foundation (NSF) and ILC Dover of Frederica, Delaware, joined forces to investigate the potential of inflatable structures as long-term habitats that may one day be deployed on the surface of Mars. In January 2008, an inflatable structure was erected at McMurdo Station, Antarctica, for the purposes of testing the integrity of the habitat in an extreme environment, analogous to the surface of Mars. The habitat (Figure 10.4), which weighs less than 500 kg, consists of a tubular inflatable structure, an insulation blanket, power and lighting systems, heaters, a pressurization system and a protective floor. While the habitat is deployed in Antarctica, engineers will have the opportunity to test its structural integrity and investigate the efficacy of dust mitigation strategies.

Although the word "inflatable" doesn't conjure up images of cutting-edge spaceflight technology, given all the advantages of inflatable structures, it is possible this habitat design will be the module of choice when astronauts eventually travel to Mars. When that time arrives, it is almost certain private companies such as Bigelow Aerospace will play a vital role in the habitat's construction and deployment.

Surface endoskeletal inflatable module

Another possible habitat structure that may be used on Mars is the Surface Endoskeletal Inflatable Module (SEIM), a concept similar to the TransHab design. The SEIM is an adaptation of the TransHab design combining the inflatable shell used for the TransHab module with a rigid frame comprising endcones, longerons, movable modular panels and one static panel. The endcones provide a means of attachment to the launch platform and serve as airlocks following deployment of the habitat, while the longerons provide the primary load resisting elements of the frame.

The construction of the inflatable shell is similar to the TransHab's, comprising an inner liner providing fire retardant protection and abrasion protection, three

Figure 10.4 Inflatable structure being tested at McMurdo Station, Antarctica. (NASA.)

bladders forming air seals in addition to four layers of felt, providing evacuation between the bladder layers. The internal pressure exerted upon the bladders is resisted by woven Kevlar straps that form the structural restraint layer. Micrometeorite, debris and radiation shielding is incorporated into the outer layer. Following deployment, the horizontal longerons act as beams, providing enough support for the installation of floors in the habitat. For a manned mission to Mars in which every gram costs a fortune, the benefits of a superlight structural system that can be easily transported and delivered are obvious, but the benefits of SEIM go beyond cost since it is designed to be restowed, redeployable and capable of supporting multiple configurations.

In-situ resource utilization

The composition of the Martian atmosphere will allow the crew to take advantage of in-situ resource utilization (ISRU) to generate propellant for the ascent stage of the Mars Surface Habitat (MSH) and the pressurized rover. Since the Martian atmosphere is composed mostly of carbon dioxide, by utilizing simple chemical reactions between Martian carbon dioxide and imported hydrogen, it will be possible to produce methane which can be used for propellant, as well as water and oxygen which can be used for life support reserves.

By the time a manned mission to Mars is undertaken, it may not even be necessary for astronauts to bring their own hydrogen as robotic missions may have detected Martian subsurface water-ice reservoirs. If the crew can find these water-ice reservoirs, they will be able to produce hydrogen through electrolysis of the melted ice.

Sabatier process

To produce methane, water and oxygen the crew will rely on the tried and trusted Sabatier process which involves the reaction of hydrogen with carbon dioxide at elevated temperatures. This reaction will be performed by the ISRU module by reacting hydrogen with Martian carbon dioxide, producing two tonnes of methane and more than four tonnes of water for each tonne of hydrogen reacted. The methane will then be cryogenically stored to be used by the ascent stage engines, which will lift the ascent stage into LMO. The water produced will be electrolyzed to produce oxygen, some of which will be used to fuel the MSH and some for life support.

Carbon dioxide electrolysis

In addition to water electrolysis, the crew will utilize another ISRU process, known as carbon dioxide electrolysis, to produce oxygen. Carbon dioxide electrolysis converts atmospheric carbon dioxide directly into oxygen by passing carbon dioxide through a zirconia electrolysis cell at eight hundred degrees Celsius, resulting in thirty percent of the carbon dioxide dissociating into oxygen and carbon monoxide.

Surface mobility

To explore beyond the local area, astronauts will require transport to assist them in their surface exploration activities. For distances beyond one kilometer but less than ten kilometers from the outpost, astronauts will use unpressurized rovers, and for regional excursions, a pressurized rover will be employed.

Robotic rovers

Exploration on a regional scale will also be performed by robotic rovers, such as the Recon Rover (Figure 10.5), which will be pre-deployed to scout out landing locations prior to the arrival of the manned missions.

In addition to exploration duties, robotic rovers will also be tasked with digging, carrying, grading, trenching, and a myriad of other construction-related activities dedicated to preparing for an extended human presence. To perform these activities, astronauts will use multi-purpose rover types, capable of performing tasks ranging from digging trenches and placing regolith, to lifting items such as the habitat module.

Pressurized rovers

Once exploration missions commence, the crew ventures further from the base and may use a pressurized platform such as the ATHLETE configuration (Figure 10.6). The ATHLETE's multi-wheeled dexterity will allow astronauts to load, transport, manipulate and deposit payloads to any site on the lunar surface, capabilities sure to make it the RV of choice.

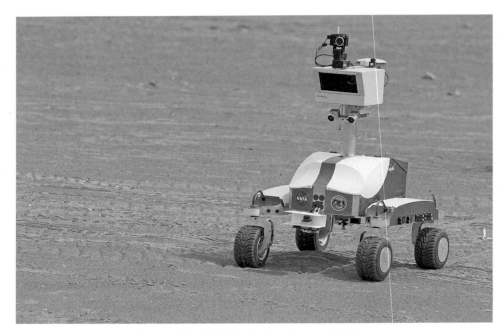

Figure 10.5 NASA's Recon Rover. (NASA.)

Figure 10.6 The All-Terrain Hex-Limbed Extra-Terrestrial Explorer (ATHLETE) will provide multi-wheeled dexterity to Martian surface operations. (NASA.)

MARTIAN COMMUNICATIONS AND NAVIGATION CAPABILITIES

Communications overview
Reliable communication systems will be critical to supporting human surface operations on Mars. In developing a planetary communication system, engineers consider the operational requirements of the mission architecture and take into account the various surface elements to support surface exploration scenarios such as those described in this chapter.

Surface-based communication network: concept of operations
Based on experience during seasons at the NASA Haughton Mars Project (HMP), the Mars analog field research program on Devon Island (see Chapter 4), Richard Alena and Stephen Braham have developed a hybrid communication architecture that may resemble the communication system ultimately adopted on the first manned mission. The hybrid architecture is a part of the Mobile Exploration System (MEX) developed at Ames Research Center (ARC) and Simon Fraser University (SFU).

Communication and navigation services
Outpost Communications and Navigation (C&N) services will be provided via relays and mobile repeaters mounted on either All-Terrain Vehicles (ATVs) controlled by astronauts, or mounted on autonomous robots deployed to optimal communication locations, thereby creating a deployable wireless network providing high bandwidth to regional and remote areas.

Mobile exploration system
The MEX architecture enables surface explorers to integrate a variety of devices and tools into their exploration infrastructure. Some of these elements are Earth-bound, some, such as the MARSAT are space-based, and elements such as the rovers, antennae and repeaters are surface-based.

As astronauts conduct surface exploration activities they record field data using wrist-mounted graphics tablets (Figure 10.7) and transmit data to base camp via Internet connectivity thanks to a single unified Transmission Control Protocol/Internet Protocol (TCP/IP)[1] subnet, located at base camp, serving all computers, both surface and space-based. This configuration permits all elements to talk to one another either directly or using TCP/IP protocols, using a satellite ground station.

Studies conducted at the HMP site, where the terrain is very similar to that on Mars, required the distance between repeater sites to be no more than three kilometers due to the topography and the requirement for a clear line-of-sight between the base and the satellite dish. To fulfill line-of-sight requirement, astronauts will position the central repeater on a high hill near base camp, providing

---

1   A TCP/IP is a suite of communications protocols used to connect hosts on the Internet.

Figure 10.7 Astronaut recording surface exploration activities using wrist-mounted graphics tablet. (NASA.)

an unobstructed view of the local exploration region. The primary repeater, located on a nearby hill, provides the height required to reach other repeater sites and, also, to reach the mobile repeaters traversing the surface or parked at exploration sites.

Mobile repeaters are hybrid rover/ATVs with directional antennae mounted on an elevated platform at the rear of the vehicle. Not only do the mobile repeaters enable continuous communication with base camp, they also serve as a platform for extending the range of the regional network, especially in areas that are out of line-of-sight of the primary repeater. The mobile repeaters also carry an array of exploration equipment such as digital/video cameras, GPS receivers, spectrometers, magnetometers and graphics tablets mounted on the handlebars, which assist the astronaut in surveying a site.

Mars communication terminals

Complementing the MEX architecture are surface Mars Communication Terminals (MCTs) with periodic Direct to Earth (DTE) capability via a Mars Aerostationary Relay Satellite (MARSAT), providing coverage of the outpost and the rest of Mars with communication, tracking and time services via the MCTs.

When embarking upon local and regional excursions, astronauts utilize either the repeaters or the MCT, a system with a 200 Mbps uplink capability to the MARSAT.

During these excursions, astronauts will also utilize cell phones and the Martian local area network (MLAN), which will support the portable Fixed Base Radio (FBR) crewmembers will use when driving the rovers/ATVs.

Mars to Earth communication: concept of operations

During surface missions, outpost communication with mission control will be routed via the MCT and a MARSAT. The combined capabilities of the MCT and the MARSAT will provide services such as forward and return voice, video, one and two-way ranging, in addition to providing fully routed data between Earth, Mars orbit, and Martian surface users, as described in Table 10.6.

Descent and landing navigation capability

Installed near the outpost will be an Autonomous Landing and Hazard Avoidance Technology (ALHAT)-based infrastructure, comprising a passive optical system and strobe lights designed to be used in the last 300m of descent in low-light landing situations. The ALHAT-based infrastructure will provide crewmembers landing on Mars with the capability to maintain an accurate trajectory throughout

Table 10.6 Outpost communication traffic model[1]

Description	System	Data Rates Low Rate (Mbps)	High Rate (Mbps)	Total Rate (Mbps)
Aggregate Peak Rate to Earth	MARSAT & Earth Ground System	**3.9**	429.0	**154.9**
Aggregate Peak Rate from Earth	MARSAT & Earth Ground System	**1.1**	66.0	**67.1**
Aggregate Peak Rate Up to MARSAT from Mars Surface	MARSAT & MCT	**6.4**	216.0	**222.4**
Aggregate Peak Rate Down from MARSAT to Mars Surface	MARSAT & MCT	**6.1**	141.0	**147.1**
Aggregate Peak Rate Across Mars Surface	MCT	**8.7**	143.0	**151.7**

Operational data transmitted to Mars (**Forward Link**): 7145-7190 MHz
Operational data transmitted from Mars (**Return Link**): 8400-8450 MHz
Mission data transmitted to Mars (**Forward Link**): 34.2-34.7 GHz
Mission data transmitted from Mars (**Return Link**): 31.8- 32.3 GHz

1. Adapted from NASA's Lunar Communications and Navigation Architecture Technology Exchange Conference, 15 Nov 2007.

the powered descent regime, and the ability to land within one meter of the intended landing site.

Surface mobility navigation capability

During their extended stay on the Martian surface, astronauts may embark upon excursions several hundred kilometers from the outpost, a farside trek placing the crew outside the communication range provided by the repeaters. To ensure crewmembers are able to find their way back to the outpost, periodic fixes of landmarks, coupled with star tracking, will need to be taken in-situ every few minutes, a process that will utilize one and two-way S-Band Doppler information from the MARSAT.

Radiometric time architecture

Time will be a crucial factor during every surface mission. To ensure astronauts are keeping to their schedules, time accuracy will be maintained by Mission Control on Earth. Mission Control will produce a Navigation Message transmitted to the MARSAT Transceiver. The MARSAT Transceiver will compare Mission Control time against the atomic time and frequency standard, before transmitting the Navigation Message on a forward link to a surface-based transponder, which will in turn demodulate the Navigation Message and forward the message to crewmembers working on the Martian surface.

Surface communication systems

The Martian communications hub will provide astronauts with wired and wireless connectivity to major elements on the surface, such as other habitats and rovers embarked upon extended-range or farside excursions. The hub will provide communications for line of sight and beyond line of sight, via the LRS. It will also provide one and two-way ranging and Doppler tracking of elements deployed on the surface. The hub's capability will extend to serving as many as fifteen simultaneous users with an aggregated bandwidth of 80 Mbps at extended ranges of at least 5.6 kilometers. This capability will be achieved using a suite of communication devices, including fixed radios, mobile radios for rovers, and EVA suit radios. In the habitat, astronauts will have powerful 100 Mbps modulators available, used to downlink and uplink large data volumes pertaining to current operations.

Surface wireless mesh networking

With several astronauts working at the outpost, the communications traffic will be served by multiple wireless links operating simultaneously. Such a wireless environment may present communication challenges to astronauts while conducting EVAs on the surface, due to the requirement of switching between different networks. On Earth, an automated mechanism for maintaining connection as one transits between networks is known as Voice Call Continuity (VCC). To serve multi-link operations on Mars, it is likely a system similar to VCC will be developed to ensure seamless handover from one element to the next.

The exploration activities and surface systems considerations described in this chapter are still in the process of being defined, evaluated and optimized with regard to performance, efficiency, reliability, weight, volume and cost-effectiveness. To achieve the optimal exploration architecture, NASA's science operations team continues to evaluate and refine surface activities that support the science and mission objectives. In tandem with the efforts of the science operations team, the systems operations team strives to provide solutions to key issues such as life support and power generation. Gradually, a feasible surface exploration scenario will evolve and efficient surface systems will be integrated to meet the objectives of that scenario. It is possible the resulting mission scenario and surface systems will resemble those described in this chapter but, perhaps, a more revolutionary approach may be adopted, as discussed in the following chapter.

REFERENCES

1. Adams, C., and Petrov, G. Variants on the TransHab Paradigm (2): The Surface Endoskeletal Inflatable Module (SEIM). Proceedings of the 35th International Conference on Environmental Systems (ICES), Society of Automotive Engineers, Warrenburg, PA, July, 2005.
2. Lange, K.E.; Lin, C.H.; Duffield, B.E., and Hanford, A.J. Advanced Life Support Requirements Document. JSC-38571C, CTSD-ADV-245C. 2003.
3. Rapp, D. Mars Life Support Systems. The International Journal of Mars Science and Exploration. Mars 2, 72–82, 2006.
4. Stafford, K.W. Advanced Life Support Systems Integration, Modeling and Analysis Reference Missions Document. Document Number CTSD-ADV-383. Crew and Thermal Systems Division, NASA Johnson Space Center, Houston, Texas. 5 November, 2001.

11

Extreme EXPeditionary Architecture

"I may say that this is the greatest factor – the way in which the expedition is equipped – the way in which every difficulty foreseen, and precautions taken for meeting and avoiding it. Victory awaits him who has everything in order – luck, people call it. Defeat is certain for him who has neglected to take the necessary precautions in time; this is called bad luck."

> From *The South Pole*, by Roald Amundsen (1872–1928), legendary Norwegian Polar explorer, who led the first Antarctic expedition to reach the South Pole on 14th December, 1911.

In his time, Roald Amundsen was first and foremost remembered as a master of planning and executing the classic polar expedition. Although Amundsen's expeditions were undertaken almost a century ago, Exploration Class missions to Mars and beyond will more closely resemble sea voyages undertaken by explorers such as Amundsen than the mission architectures used for the Space Shuttle and International Space Station (ISS). The first voyagers to Mars will spend up to six months getting there, almost two years exploring the surface and up to six months on the return journey. To maximize their two-year stay on the surface it will be necessary for Mars crews to adopt the strategies employed by Amundsen and his fellow polar explorers, the most important of which will be mobility. In common with early polar exploration expeditions, the key to accomplishing the many scientific and exploration objectives on the Martian surface will lie in adaptable and mobile exploration architecture. Such an architecture, known as Extreme EXPeditionary Architecture (EXP-Arch), was conceived and developed by a small team of designers, engineers and scientists working for the NASA Institute for Advanced Concepts (NIAC).

MOBILE, ADAPTABLE SYSTEMS FOR MARS EXPLORATION

EXP-Arch, devised by Trotti and Associates, Inc. (TAI), represents a radical departure from conventional mission planning such as NASA's current approach, which requires an incremental build approach to establish an outpost [2]. In contrast to NASA's plan, EXP-Arch is an adaptive and versatile exploration architecture, by virtue of its utilization of multi-functional, modular and transforming-system capabilities. Although EXP-Arch retains the goals mandated by NASA's Exploration Systems Mission Directorate (ESMD), the architecture itself is nothing less than revolutionary in concept.

Revolutionary exploration objectives
In stark contrast to the conventional planetary surface architectures consisting of stationary base designs, EXP-Arch represents a paradigm shift by creating a suite of systems that are not only highly mobile but also self shielding and quickly deployable. Some of the EXP-Arch systems envisioned by Trotti and his team enabling such an architecture include mobile habitats, rovers and laboratories with the capability of transforming from one cubic meter when stowed, to more than nine cubic meters when fully deployed.

Extreme EXPeditionary Architecture background
The concept of EXP-Arch's vision of planetary exploration has its roots in the way humans have explored the Earth. Historically, exploration objectives and discoveries have been realized thanks to highly mobile systems. Amundsen used dogs to reach the South Pole in 1911, present-day Antarctic research stations use mobile laboratories and to accomplish the many exploration goals of a manned mission to the Red Planet it will also be necessary for crewmembers to be highly mobile.

In anticipation of future developments in intelligent materials and robotics, the EXP-Arch goal was to devise an exploration strategy with a high degree of autonomy. The vision of Trotti and his colleagues was to create an architectural exploration system operating as a series of mobile vehicles capable of transporting themselves to exploration locations and then transforming themselves into a habitat and/or research station. Some of the concepts considered initially included the development of inflatable structures utilizing intelligent lightweight materials capable of self-adaption. The structures would be capable of rapid self-deployment and contain life support facilities and research equipment in addition to being able to respond to the environment.

The concept of the EXP-Arch gradually evolved during the process of researching present-day rovers, exploration systems and developments in the field of fuel cells, biomimetics and navigation systems. In this research it was determined no habitable rovers had ever been designed to operate autonomously or without the dependence on a base for re-supply. The EXP-Arch team decided for humans to truly become a space-faring species it would be necessary to provide crews with more than just a single outpost and a dependent rover. To that end, the team embarked upon devising an alternative vision with the intent of providing future Mars crews with structures and systems capable of not only self-building, self-repairing and self-propelling, but also of evolving and rapidly deploying. An example of one of the versatile systems devised by the EXP-Arch team is the pressurized Mother Ship Rover (MSR, Figure 11.1), a pressurized vehicle capable of transporting and supporting four crewmembers during forays across the Martian surface.

Figure 11.1 EXP-Arch Mother Ship Rover/Scorpion. Design: Trotti Studio. (Mitchell Joachim.)

EXP-ARCH VEHICLE DESIGN CONCEPTS

Design principles

Two integral components of the EXP-Arch mission profile are the aforementioned MSR and Mini-Rover, the details of which are summarized in Table 11.1.

Initial conceptual designs

The MSR and mini-rover designs may appear vaguely familiar to those who have studied marine biology due to the novel approach adopted by the EXP-Arch team in examining ways nature solves problems of adapting to environments. For example, *Lagocephaslus lagocephalus*, more commonly known as the puffer fish, maintains a hydrodynamic shape while moving, but as a defense mechanism puffers have the ability to inflate rapidly, filling their extremely elastic stomachs with water until they are almost spherical. The advantages of such unique characteristics were not lost on the EXP-Arch team, who decided to incorporate the idea into the design of their inflatable habitats and airlocks.

Inspiration also came from studying the armadillo as a potential model for micrometeorite shielding, while the scorpion served as a baseline for the design of the Gamma Rover.

Table 11.1 Mother Ship Rover and Mini-Rover general and specific design criteria

MSR and Mini Rover General Design Criteria

All terrain omni-directional vehicle	Modular design of lightest possible structure.	Night Vision and video cameras.
Topography detectors and sensors.	Teleoperation.	Navigational warning systems.
Highest degree of autonomy.	Low dust pick-up wheels.	High traction.
Independent wheels.	GPS Martian navigation.	Solar and fuel cells.

MSR Specific Design Criteria

Experimental airlock/glove box.	'Dirt room' for space suit storage and repair.	Wardroom dining & videoconferencing.
Micrometeorite shielding.	Hygiene area.	Galley.
Re-supply of fuel, atmosphere, food and water.	Maximum speed 15 kmh. Maximum mass, 18 mT.	Crew of four.
Sleeping area for all crew.	Expandable environment.	70 KPa pressure.

Mini Rover Specific Criteria

Partial micrometeorite shielding.	Non-pressurized.	Modular design.
Crew of four in EVA suits.	Maximum speed 20 kmh.	2 mT load capacity.

Following the first phase of the design process it was decided the pressurized MSR and unpressurized mini-rover would serve as the primary vehicle classes from which variants would be derived. The next phase of the design process involved conceptualizing design alternatives and subsystems such as the Beta Rover and Delta MSR.

Final vehicle design: MSR Scorpion
Following the conceptualization phase, the EXP-Arch team moved on to finalizing vehicle designs. The primary class of rover, designated the 'Scorpion'' (Table 11.2), is a highly mobile autonomous rover whose exterior characteristics pass more than a passing resemblance to the creature inspiring its design (Figure 11.2).

MSR Scorpion exterior design
Although the design of the MSR Scorpion may appear to have been the product of the fevered imagination of a science fiction artist, all the features perform functions specific to dealing with the hostile Martian environment. For example, the arrays radiating away from the Scorpion's hull are attached to telescopic arms mounted on three-degrees-of freedom joints, enabling the arrays to track the weak Martian sunlight, thereby maximizing power output. Another critical feature is the meteorite

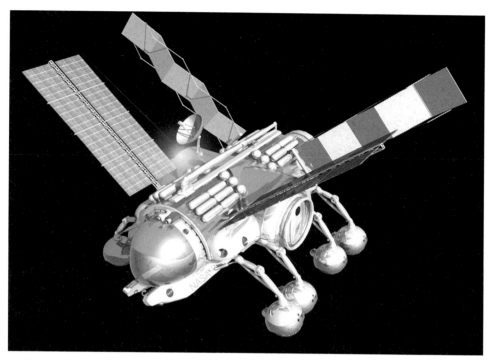

Figure 11.2 EXP-Arch Mother Ship Rover/Scorpion. Side view. Design: Trotti Studio. (Mitchell Joachim.) See Plate 14 in color section.

Table 11.2 Mother Ship Rover "Scorpion" specifications

Pressurized Structure Dimensions	
Length	11.2 m
Width	5.8 m
Height	3.4 m
Weight	12 mT
Internal Volume	65-80 m^3
Solar Array, Fuel Cells and Power Features	
Solar Array Power (each)	50 kW
Solar Array Area	75 m^2
Deployed Pressurized Inflatable Environment Dimensions	
Length	9.15 m
Diameter	3.04 m
Deployed Pressurized Inflatable Airlock	
Length	2.44 m
Diameter	3.04 m

shield positioned 25 centimeters away from the pressurized hull. Constructed of ceramics and the very latest nanotechnology materials, the Scorpion's meteorite shield, by virtue of its location on the hull, forms a robust defense against high angle impacts.

MSR drive train and underside features
Like its eight-legged arthropod namesake, the Scorpion features eight articulated legs with four degrees-of-freedom, capable of raising the pressurized hull two meters above the Martian surface. The mother ship is also fitted with novel spherical gridded wheels one meter in diameter. Constructed of an elastic mesh grid, the wheels also function as a suspension system and, thanks to their spherical design, are capable of maximizing contact with the surface at various angles.

The two prominent skids under the Scorpion protect the hull when the vehicle sits on the surface and if it becomes stuck in the sand of one of the Martian deserts. In such an event the Scorpion is capable of self-rescue by deploying winches and pull-up hooks mounted between and on the skids.

MSR Scorpion interior design
The Scorpion was designed to provide a crew of four with a productive working and living environment (Figure 11.3). Crewmembers ingress the vehicle via the airlock located at the rear and enter the EVA storage and repair compartment, notable for the four EVA suits hanging on the bulkheads. The next section is the hygiene compartment, followed by the Storm Shelter Hatch (SSH), located between the

Figure 11.3 Scorpion interior. Design: Trotti Studio. (Mitchell Joachim.)

inflatable environment and the food storage racks. The modular and exchangeable food storage racks will be familiar to anyone familiar with the interior of the ISS. Once astronauts step over the SSH, they enter the wardroom and galley, located at the centre of the vehicle. Workstations are fitted on the port and starboard bulkheads equipped with all the necessary computer facilities. Designed utilizing the very latest developments in aircraft human factors engineering, the workstations feature ergonomic seats, computer hook-ups and small portholes providing crewmembers with alien vistas as they work. Mounted on a platform in a glass dome, flanked by two fully reclining seats/beds is the cockpit, the nerve centre of the vehicle.

Final vehicle design: Mini Rover

In common with the design of the MSR, the Mini Rover (Table 11.3) is capable of being operated autonomously, by tele-operation and by a human pilot. Highly mobile, agile and tremendously versatile [3], the Mini Rovers are capable of transporting cargo, sample return, moving regolith and serving as storm shelters in the event of a solar particle event (SPE).

The support platform for the drive train, fuel cells and communication system is based on a model of a skateboard, a concept currently being developed by North American automotive manufacturer, General Motors. The skateboard concept employs an open-design principle, featuring coupling points permitting modules such as Power, Safety, Engines and Information Systems to be attached to the platform. Once attached, the modules run independently and permit operators to swap parts as required, easily transforming a sample return rover into a construction vehicle.

The Mini Rovers are intended to be the workhorses of Martian expeditionary missions, a role reflected by the equipment they carry. In addition to winches, trench diggers, auger guns and drills, the Mini Rover is also equipped with a highly dexterous manipulator arm capable of loading and unloading cargo racks and collecting samples.

Table 11.3 Mini Rover specifications

Mini Rover Features	
All Terrain Vehicle	Four wheel drive
Operational during day and night cycles.	Crew transports for four astronauts.
Stereoscopic Vision	Wireless Controls
Adaptable to mission requirements	Tele-operation capable
Mini Rover Capabilities	
Cargo Transport.	MSR Assistant
Cargo Transport	Regolith Mover.

RADIATION AND SHELTER STRATEGY

Chapter 8 described the dangers of radiation and the crucial requirement of protecting crews from the potentially lethal effects of galactic cosmic rays (GCRs) and solar particle events (SPEs).

In common with the idea for the design of the MSR and Mini Rovers, the radiation protection strategy adopted by EXP-Arch was bio-inspired. In devising a radiation protection strategy the EXP-Arch team studied defense mechanisms adopted by creatures such as hermit crabs, clams and snakes. Most species of hermit crabs are protected from predators by carrying around a seashell, into which the whole crab's body can retract. Freshwater clams, by comparison, dig themselves into sand to protect themselves against any potential predators. Although astronauts on the surface of Mars will hopefully not have to protect themselves from predators, it is possible the methods adopted by crabs and clams will be utilized to provide protection against radiation.

The pressurized rovers designed by the EXP-Arch team will carry high density equipment in addition to water tanks and batteries on the roof and sides of the vehicles. Although this shielding may protect crews from nominal radiation levels encountered on the surface of Mars it is unlikely it will afford sufficient protection in the event of an SPE. In such an event, crews will protect themselves by adopting the defense strategy of the freshwater clam and create a 'deep well' shelter. Since a solar flare cannot be predicted, crewmembers will not have much time to build their 'deep well' since the X-rays and gamma rays emitted by an SPE travel at the speed of light and may reach Mars in as little as thirteen minutes, although the most energetic flare particles will arrive between three to fifteen hours after the outburst. Given these timelines (Table 11.4) it will be necessary for crews to utilize custom-designed explosives, unless there happens to be a suitably shaped crater nearby.

By using the explosives the crew will attempt to create a crater six meters deep and ten meters in diameter and then simply maneuver the Scorpion over the crater.

ADVANCED COMPOSITE MATERIALS

Space travel has produced a number of textiles that have found their way into everyday use. For example, Kevlar and Gore-Tex were both developed by NASA and later marketed by DuPont, and it seems the tradition of developing innovative textiles will continue in realizing the adaptable, mobile architecture envisioned by the EXP-Arch team.

Thermo-sets
Many of the concepts proposed by the EXP-Arch team are based on advanced lightweight structures consisting of high performance resin systems with fibre reinforcement. The resins may be divided into two categories; *thermosets* and *thermo-plastics*.

Table 11.4 Radiation shelter deployment

Elapsed Time h m s	Event	Crew Action
00:00:00	*Solar Flare Alarm*	Searches for site to create a shelter.
00:15:00	*Unmanned mini Rover deployed to shelter site to drill and set explosives.*	Tele-operates Mini Rover to site of shelter.
01:00:00	*Explosive charge is placed at shelter site.*	Mini Rover tele-operated away from shelter site.
01:01:00	*Explosive charge detonated.*	Prepares to travel to shelter site.
01:30:00	*Travel to shelter site.*	Deploys Mini Rovers to install tools to cover shelter.
01:45:00	*Inflatable shelter deployed.*	Prepares equipment required to survive storm.
02:00:00	*Emergency equipment deployed.*	Tele-operates Mini Rovers to cover shelter.
02:30:00	*Crew Ingress shelter.*	Monitors radiation levels inside and outside shelter.
96:00:00	*End of flare.*	Empties, evacuates and depressurizes shelter.

Thermosets, also known as thermosetting plastics, are polymer materials that irreversibly cure to form a stronger material. The process of curing is usually achieved by heating the material to above 200°C or using irradiation such as electron beam processing. During the curing process, energy is added causing molecular chains to react at chemically active sites, eventually linking into a rigid, three-dimensional structure. In addition to creating three-dimensional bonds, the curing process transforms the resin into a plastic by a cross-linking process forming a molecule with a larger molecular weight, resulting in a material with a higher melting point. During the reaction, once the molecular weight has increased to the threshold at which the melting point is higher than the ambient temperature, the material forms a solid material. As a result of the three-dimensional network of bonds and the curing reaction, thermo-set materials are not only very resistant to heat but also exceptionally strong, making them ideal candidates for constructing Martian rovers.

The structural and heat-resistant characteristics of thermo-sets may be different depending on whether the material is amorphous or semi-crystalline. An amorphous high temperature resin has a randomly ordered structure which does not have a sharp melt point but instead softens gradually as the temperature rises beyond its glass transition temperature. Since the glass transition temperature is normally above 200°C, this is not a concern for Mars where the average surface temperature is –61°C and the maximum temperature is 20°C. In contrast, semi-crystalline materials have a highly ordered molecular structure with sharp melt points and do not gradually soften with temperature increase. This means they retain their strength, especially when reinforced with other materials such a fiber.

Three-dimensional braided fabrics

A braid is a complex structure formed by intertwining three or more strands of flexible material such as textile fibers [6]. The braiding of such fibers is an approach utilized by EXP-Arch to manufacture the composites required to construct the vehicles and structures. Because the process offers the optimum combination of conformability, torsional stability and structural integrity, braided structures have already been utilized in the area of rocket engine nozzles and nose cones. Given braiding techniques are also capable of producing highly intricate designs, the process is ideally suited to realizing many elements of the proposed EXP-Arch vision.

BIOINSPIRED ENGINEERING OF EXPLORATION SYSTEMS

A manned Mars mission will be a sophisticated expedition, demanding the greatest complexity and functionality of any mission ever undertaken. Although astronauts will spend a large proportion of their time conducting exploration activities, a significant amount of their workday will also be spent attending to life support systems, repairing equipment and performing routine maintenance tasks. Ideally then, to ensure the science return is maximized it will be helpful if the astronauts could deploy autonomous explorers, capable of surveying the surface of Mars, conducting science experiments and performing scouting missions. Fortunately, the means to achieve this goal is available thanks to a multidisciplinary concept known as bioinspired engineering of exploration systems (BEES).

> "How biological systems store and retrieve information, control development, fabricate structural components, build molecular machines, sense the external environment, reproduce and disperse themselves throughout the environment, engage in error detection, and carry out self-repair can pay big dividends to space exploration." [1]

BEES represents a new approach NASA is adopting to space exploration. By studying how living things interact with their environment, the agency plans to apply similar principles to explore planets such as Mars. However, the intent is not just to mimic operational mechanisms found in a specific biological organism but to take the best features from a variety of diverse bio-organisms and apply them to a desired exploration function. Adopting this approach, engineers hope to build explorer systems with specific capabilities endowed beyond nature as the systems will possess a combination of the very best nature-tested mechanisms for a specific function. Such an approach is a logical one since, through billions of years of evolution, nature has perfected its designs, and by selecting the best of these designs it will be possible to go beyond biology and achieve unprecedented capability and adaptability in explorer systems.

Already, work is underway to develop what engineers call *biomorphic explorers*, using principles of a BEES discipline known as *biomimetics*, which deals systematically with the technical execution and implementation of construction processes and developmental principles of biological systems. Using biomimetic principles, engineers plan to develop the classification of biomorphic explorers into two main classes of surface and aerial explorers, examples of which are described here.

Gecko-tech

Geckos
Thanks to a mysterious gecko adhesive, geckos have the ability to walk up walls and run across ceilings, skills that could be usefully employed by astronauts exploring the surface of Mars. Wearing a Gecko-Tech biomimetic suit, astronauts could achieve extraordinary mobility capabilities leaving few places on the Red Planet inaccessible.

The key to the gecko's mobility is achieved thanks to evenly-spaced gripping strips crossing the end of the gecko's toe. These strips, also known as *lamellae*, are comprised of special hair-like structures called *setae*. On each foot there are about 500,000 setae, each seta consisting of tiny pad-on-stem structures called *spatulae*, which act as the gripping structure enabling the gecko to perform its acrobatic mobility. Research investigating materials possessing the necessary mechanical tensile strength, flexibility and formability suitable for fabricating artificial spatulae and setae, suggest gecko-tech adhesion is a realizable goal using nanometer-scale fabrication techniques. Gecko-derived spacesuits would enable astronauts on Mars to scale the walls of the Martian valleys and explore almost any terrain by being able to climb over any obstacles, thereby saving time over a more conservative 'going around' route. Using the gecko-suit, designed with gecko-style adhesive pads on the hands, knees and feet, Martian explorers would not only be able to climb vertically but also achieve inverted mobility!

Entomopters
During the 1990s the Defense Advanced Research Project (DARPA) considered the feasibility of using minute flying vehicles on the scale of insects, a line of research that gave birth to the *'Entomopter'*, a descriptor combining the concept of entomology with the word helicopter. It wasn't long before NASA, an agency familiar with the potential of airborne technologies to conduct exploration, decided to pursue its own research.

Flying on Mars
One of the problems of flying on Mars is the rarefied atmosphere requiring aircraft to fly within a very low Reynolds number (see Glossary) and high Mach number regime. This means an aircraft must have a large surface area and fly at very high speeds to generate sufficient lift, which isn't very helpful for exploring. Another means of achieving the required lift is to flap the wings, but simply flapping up and down is still not sufficient to achieve flight in the Mars environment. To produce sufficient lift, the flapping must be combined with an additional lift-producing

mechanism known as the Magnus force, a rotational motion insects use when flapping their wings. By replicating these insect flying mechanisms, researchers developed the Entomopter, capable of flight in the Martian environment and able to operate close to the surface.

Entomopter hardware

One of the first considerations addressed by Entomopter researchers was the issue of how to provide the energy to propel the biologically inspired wings. Batteries couldn't be used because of the weight, so scientists developed a device called a Reciprocating Chemical Muscle (RCM), an ignition-free, catalytic contraption capable of operating from a number of chemical fuel sources and one that did not require any oxygen.

The RCM works by converting chemical energy into motion through a direct non-combustive chemical reaction, hence the term 'muscle', as opposed to 'engine'. Not only does the RCM enable automatic wing-flapping, the mechanism also produces small amounts of electricity sufficient to power onboard reconnaissance devices, thereby enabling exploration.

The next problem scientists had to solve was how best to fly in the Martian atmosphere. By examining wing-flapping frequencies and testing various designs in the Jet Propulsion Laboratory's (JPL) Mars atmosphere simulation chamber, scientists were able to define the optimum balance of flapping and the previously mentioned rotational Magnus effect.

The next stage in the Entomopter design was to develop the aerodynamic shape of the vehicle, a challenge complicated by the characteristics of the tenuous Martian atmosphere and a lack of data and tools concerning simulation of the complex kinematics involved in insect flight. The design chosen was an X-wing configuration with two sets of wings. Unlike Nature's way of flapping, which requires the wings on both sides to be flapping in the same direction, the X-wing version is one in which one side flaps up and the other side flaps down and vice versa.

Concept of operations

The Mars Entomopter may be used by crewmembers to scout out potential locations of interest and to conduct surveying tasks. A baseline mission, which will be highly if not completely autonomous, will begin with crewmembers programming the days tasks into the Entomopter's computer. Once the Entomopter is programmed it will launch from its refueling rover and begin its tasks. Although the Entomopter's flights will only last between five and ten minutes, the flights will be numerous thanks to the refueling rover.

To avoid crewmembers having to monitor the Mars Entomopter, it will be fitted with flight control, navigation and collision avoidance software just like a regular aircraft. A rover-centric system will also be a part of the Entomopter's software, providing updates of its relative position to the refueling rover so it doesn't run out of fuel. Also in common with a regular aircraft, the Entomopter will be fitted with radar, enabling it to not only map obstacles, but also provide it with an inherent homing beacon capability and bidirectional communication via direct transmission from the refueling rover.

Biomorphic explorers

Biomorphic explorers will conduct Biomorphic missions characterized by synergistic use of existing and conventional surface and aerial assets such as the habitat, rover and orbiting MTV. The types of science objectives assigned to this new class of autonomous explorers include close-up imaging for identifying hazards, assessing geological sites, gathering atmospheric information and deploying surface payloads such as instruments or surface experiments [4, 5]. To achieve these objectives, these reconfigurable biomorphic explorers will be programmed by crew for specific functions.

Sample Biomorphic missions: imaging and site selection

One of the first exploration objectives of the crew will be to survey and select sites of interest. Although imaging from orbiting vehicles allows broad coverage the spatial resolution is limited to only one meter. To save time, the astronauts will find it useful to have accurate imaging of the local area so they can plan exploration paths for the rover and also characterize sample return sites. Ideally, to achieve this goal, coverage of a large area is required and close-up imaging in the five to ten centimeter ranges is desirable so that hazards and slopes can be identified for a successful rover mission. To achieve these goals bioflyers, equipped with sensors and miniature cameras, will be deployed on a variety of flight paths to image the horizon terrain. If a bioflyer identifies a potential exobiological site it makes decision to terminate the flight and deploys a small science experiment with a pyrotechnic device that disengages from the bioflyer and penetrates the ground. Meanwhile, the bioflyer orbits the exobiological site waiting for data to be transmitted by the science payload. Once the data is received, the bioflyer transmits the results of tests to the habitat and, if required, continues to orbit the site to serve as a beacon for the crew. The crew then begin their mission, driving to the site identified by the bioflyer. Thanks to the imaging information sent by the bioflyer, the crew are able to take the most direct and safest route to the site, thereby saving time and reducing risk.

Sample Biomorphic missions: surface experiments

In this mission, a biomorphic orbiter is deployed to an area of geological or scientific interest. Onboard the orbiters are ten to twenty seed wing flyers, each equipped with a small surface probe, chemical experiment payload and a miniature camera. The orbiter traverses the area of interest while simultaneously gathering meteorological data such as weather patterns used to select the timing of release of the seed wing flyers. Once an area of interest is identified, seed wing flyers are released and land on the surface, where they conduct surface experiments that may include testing for trace elements and/or chemical testing. Once the science experiment has been completed the seed wing flyer transmits its data to the orbiter, which in turn transmits the data to the habitat where the astronauts can analyze it.

Sample Biomorphic missions: aerial reconnaissance

There may be occasions when the orbiter information relayed to the astronauts in the habitat is just not sufficiently detailed to make an assessment of large scale geological or meteorological data. In such an event, the astronauts will release a squadron of small biomorphic gliders equipped with small infrared (IR) cameras and surface probes. The gliders are pre-programmed by the astronauts with the coordinates provided by the orbiter and deployed to priority targets. As the gliders fly over their targets they transmit high resolution imagery back to the astronauts sitting in the habitat. From the habitat the astronauts can command the gliders to take different flight paths in order to image the area from different aspect angles and also instruct the gliders to land if a particular area looks like it may yield valuable data.

Sample Biomorphic missions: local and regional sample return
In this mission, the objective is to obtain samples from potential exobiology sites and areas of geological interest. To do this, the astronauts deploy an autonomous rover loaded with an arsenal of scientific experiments and a squadron of biomorphic explorers equipped with a miniature camera and a small IR detector. Once at the target location, the rover deploys the scientific experiments and transmits the results to the astronauts back in the habitat. If the results appear interesting, the astronauts command the biomorphic explorers to return samples while the rover continues on its mission.

Sample Biomorphic missions: deployment to the polar ice cap
The habitat is located a great distance from the polar ice cap but given the potential of the polar site to yield invaluable data concerning the historical climatology of Mars it is inevitable scientists will want to perform science there. Once again, the astronauts will program an autonomous rover to travel to the ice cap loaded with a science package and several biomorphic explorers designed to burrow through ice, snow and soil using a combination of scraping and heat while pulling debris around itself and applying downward pressure with its limbs. Once on the ice cap, the burrowing biomorphic explorers would do their job but progress would be slow due to the limited light available for solar power. To ensure power supplies are not depleted the biomorphic explorers would need to stop their burrowing occasionally to enable solar cells to recharge their batteries. As they make their way through the ice, instrumentation would detect different ice layers and determine the composition, data that would be transmitted via the rover back to the habitat.

Yabbies
Another biomimetic application that may assist crewmembers is an engineered version of the common freshwater crayfish, known in Australia as the *Yabby*. At the University of Melbourne, scientists and zoologists think crews on Mars will be able to deploy teams of *robo-yabbies* to help search for water or conduct chemical analysis of planet's crust, tasks that may be impractical or dangerous for humans.

Thanks to advances in computational network modeling (CNM), engineers have managed to apply the movement of the Yabby's tail to the design of robots capable of traversing over difficult terrain. The Yabby's segmented tail, which acts like a hinged lever capable of changing to act as a sail for steering or as a paddle for swimming, gave scientists the idea of using a similar system in the design of miniature, lightweight robots with multi-jointed legs, capable of performing a range of complex tasks needed to explore Mars.

Whichever robots astronauts decide to take with them to Mars, the chances are that many of them will be biologically inspired since the advantages of a biomimetic robot have several benefits over traditional robots. Not only can biomimetic robots provide explorers with a greater range of terrain-crossing capabilities, the increased autonomous control and learning behaviors will increase the scientific rewards of the missions as time delays with Earth-based decision-making will be removed from the equation.

REFERENCES

1. Mjolsness, E., and Tavormina, A. The Synergy of Biology, Intelligent Systems and Space Exploration. IEEE Intelligent Systems 15 (2): 20–25. 2000.
2. NASA. NASA's Exploration Systems Architecture Study (ESAS). From http://www.nasa.gov/mission_pages/exploration/news/ESAS_report.html 2005.
3. Saleh, J.H.; Hastings, D.E.; and Newman, D.J. Flexibility in System design and Implications for Aerospace Systems. Acta Astronautica. Vol. 53, pp. 927–944, 2003.
4. Thakoor, S. 1st NASA/JPL Workshop on Biomorphic Explorers for Future Missions. NASA's Jet Propulsion Laboratory Auditorium. 4800 Oak Grove Drive, Pasadena, CA 91109. August 19–20, 1998.
4. Thakoor, S.; Miralles, C.; Martin, T.; Kahn, R.; Zurek, R. Cooperative Mission Concepts Using Biomorphic Explorers. Jet Propulsion Laboratory, 4800 Oak Grove Drive, Pasadena, CA 91109. Lunar and Planetary Science XXX. 1999.
5. Wang, Y.; Zhao, D. Effect of Fabric Structures on the Mechanical Properties of 3-D Textile Composites. Journal of Industrial Textiles, 35 (3): 239–256. 2006.

Epilogue

The United States is a country at war. A war that is a hemorrhage of money, resources and national morale. It is also a war that has that has taken center stage while the dialogue for resurrecting the manned space program has petered out. In the aftermath of the Columbia tragedy the question of manned spaceflight took an even bigger hit when the naysayers suggested that the entire manned space program was not worth the risk. The American public still say they want to send people into space, but it's been a generation since they've felt strongly about it and even longer since they've been inspired by what astronauts do. The astronauts of today are no longer the stars that Neil Armstrong and Buzz Aldrin were. As brave and bold as it may be, launching into low Earth orbit (LEO) to rendezvous and dock with the ISS and perform science experiments just doesn't inspire today's generation, but without a Moon race driven by a cold war there is nothing else for astronauts to do. Thanks to the Pentagon, which has been extraordinarily transparent about what the public know of the cost of the war, we know that the monthly cost of maintaining the war in Iraq exceeds twelve billion dollars a month. It may seem like an exaggerated number, but it isn't. Instead of going to Iraq, the United States could have funded an independent manned Mars mission every four months. Of course, this brings us onto the subject of cost and to two contrasting arguments that just never go away when discussing the space program. First, there are the "it's a waste of money" types who argue we'd be better off spending the money solving problems on Earth and then there are the "Mars or bust" types who have spent too much time reading Ben Bova and Kim Stanley Robinson novels who argue that we've been trapped in LEO for far too long and we must fly to Mars whatever the cost.

Earth's close planetary neighbor has always been a source of fascination by offering us a tantalizing glimpse of what the future might hold and the promises of what will be our first steps onto another world. Travelling to Mars will undoubtedly be a high risk venture and even greater risks are sure to follow when humans embark upon interplanetary missions to the outer planets and beyond. Astronauts and cosmonauts are well aware of the dangers spaceflight presents but like the early polar explorers they acknowledge the risks they face with each mission in the knowledge that throughout history, visionaries and explorers have been willing to take up such challenges. Achieving the interplanetary goal of landing humans on Mars and bringing them back alive will be more than just another space accomplishment. It will finally, after far too long, reaffirm the pioneering spirit of human society, an aspect that many believe has diminished since the Moon landings more than thirty-five years ago.

Those in charge of NASA know the public yearn for something more enduring and something more inspirational than sending astronauts to LEO, but in an agency stifled by its own safety culture, mounting an expedition to Mars before solving the technology issues is an unlikely proposition. An old adage in the space business states that once you've expended the energy to get into LEO, you're halfway to anywhere else in the solar system. Perhaps NASA will go the rest of the way soon.

Glossary

Earth Orbit Rendezvous

Earth Orbit Rendezvous (EOR) is a mission architecture component requiring a docking of two spacecraft in LEO. For example, the Constellation mission architecture will require unmanned mission elements to be launched on a cargo vehicle and the manned elements to be launched on the crew LV. In LEO the unmanned and manned elements will rendezvous and dock before proceeding to Mars.

Escape Velocity

For a spacecraft to reach Mars it must attain an escape velocity of 11.2 kilometers per second, which is the velocity required to escape Earth's orbit. Normally, for an actual escape orbit, a spacecraft is first placed into LEO and then accelerated at that altitude.

Human-Rating Requirements

The process of human-rating a launch vehicle is governed by NASA's Office of Safety and Mission Assurance publication, NPR 8705.2. NPR 8705.2 is also known as the NASA Human-Rating Requirements Compliance Verification Guide, a technical document designed to provide the maximum reasonable assurance of safety for people involved with spaceflight activities.

NPR 8705.2 Criteria

A Human-Rating Plan must include:	The Human-Rating Board consists of:
Design criteria.	Associate Administrator for Space Operations.
System designs.	Associate Administrator for Space Explorations.
Test requirements and procedures.	Chief Safety and Mission Assurance Officer,
Software design.	Office of Primary Responsibility.
Test and verification requirements.	Chief Medical and Health Officer, Independent
System safety and reliability engineering.	Technical Authority.
Human factors engineering requirements.	Chief Engineer, Independent Technical
Health requirements.	Authority.

NPR 8705.2 NASA Human-Rating Requirements. Compliance Verification Guide. NASA Office of Safety and Mission Assurance. Washington D.C. August 29, 2005.

NPR 8705.2 covers the spectrum of capabilities and requirements pertaining to human-rating a LV and serves as a rigorous safety mechanism ensuring spaceflight systems are certified in compliance with acceptable risk-management parameters.

Inclination, Periapsis and Apoapsis

An orbiting spacecraft will be oriented at a certain *inclination* relative to the Earth. For example, an inclination of zero degrees (also termed a low inclination orbit) indicates the spacecraft is in orbit about the Earth's equator whereas a spacecraft that has an inclination of ninety degrees (this would be termed a high inclination orbit) indicates it is in orbit about the Earth's poles. Regardless of the type of inclination, an orbiting spacecraft will follow an oval shaped path known as an *ellipse*, which means there will be a point of closest approach to Earth, known as *periapsis*, and a point farthest away from Earth, known as *apoapsis*.

Interplanetary Trajectory and Transfer Orbits

To send a spacecraft to Mars it isn't sufficient to achieve escape velocity and simply point at the Red Planet. The spacecraft must be launched on a trajectory/transfer orbit that will intersect Mars. Due to the requirement of considering the gravitational interactions of not only the Earth the Moon, Mars, but the Sun as well, the complexities associated with calculating transfer orbits often appear computationally overwhelming.

Liquid Propellants

A liquid propellant system is characterized by storing the fuel and oxidizer in separate tanks from which a series of valves, pipes, and turbopumps feed the fuel and oxidizer to the combustion chamber where they burn to produce thrust. One advantage of a liquid propellant system is its throttleable capability, achieved by controlling the propellant flow to the combustion chamber. This design characteristic also permits stopping and restarting the engine.

Low Earth Orbit

To attain Earth orbit, a spacecraft must be launched to an altitude above the Earth's atmosphere and accelerated to orbital velocity, which is between 6.9 and 7.8 kilometers per second. The inclination of the orbit desired will have a bearing upon the amount of propellant required, with low inclination orbits being the most energy efficient and high inclination orbits demanding the most propellant.

To reach a low inclination/energy efficient orbit a spacecraft is launched in an eastward direction from a site as close as possible to the equator. Such a direction and location confers an advantage on the launching spacecraft thanks to the contribution the rotational speed of the earth makes to the final orbital speed of the spacecraft. This is one reason NASA's primary launch site, Cape Canaveral, Florida, is located where it is, since the relatively low latitude of 28.5^{0}N results in a free ride equating to adding 1,471 km/h to any eastward launched orbital spacecraft.

Due to the rapid orbital decay of objects in orbits below 200km, the accepted definition of low Earth orbit (LEO) is between 160 and 2000 km above the Earth's surface. For example, the International Space Station (ISS) orbits the Earth in a LEO varying between 319.6 km and 346.9 km altitude.

Orbit Perturbations
The launch of a spacecraft comprises a period of powered flight commencing with lift-off from the launch pad and concluding upon burnout of the rocket's last stage when the vehicle is in orbit. Once in orbit the spacecraft is considered to be in a state of free flight and its trajectory is subject only to the gravitational pull of the Earth. If the spacecraft travels far from the Earth, its trajectory may be affected by the gravitational influence of the Moon or other planets, an effect referred to as *third-body orbit perturbations*. Other orbit perturbations affecting the spacecraft include *drag*.

Orbital Maneuvers
In addition to permitting spacecraft to perform altitude changes, orbital maneuvers are used to transfer from one type of orbit to another, to permit rendezvous with other spacecraft, and to change the parameters of an exiting orbit to meet a launch window. Perhaps the most complicated and most precise orbital maneuver is orbital rendezvous, which requires a spacecraft to intercept another spacecraft or object at a rendezvous point at the same time.

Orbital Mechanics
The motions of satellites and spacecraft moving under the influence of gravity, atmospheric drag, and thrust, are described by orbital mechanics, the root of which lies in Newton's law of universal gravitation. Typical applications of orbital mechanics include the calculation of ascent trajectories, re-entry and landing, orbital rendezvous, and interplanetary trajectories.

Reynolds Number
In fluid mechanics, a criterion of whether fluid or gas is absolutely streamlined is determined by the Reynolds number. A value of less than 2,000 indicates flow is streamlined whereas a value of more than 2,000 indicates flow is turbulent. The transition between streamlined and turbulent flow occurs between 1,000 and 2,000.

Rocket Propulsion.
The principle of rocket operation is described by Newton's third law of motion that states "for every action there is an equal and opposite reaction". Propellants are delivered to the combustion chamber where they react chemically to generate hot gases which are accelerated and ejected at high velocity through a nozzle. The forceful expulsion of hot gases through the nozzle provides momentum or the thrust force of the rocket motor in a similar manner to a gun recoiling after being fired.

Space Architecture
When mission planners discuss *space architecture* they are referring to the three components comprising a mission design. These components are the space segment, the launch segment and the ground segment.

Space Segment
The space segment includes a description of the orbital mechanics required to place a spacecraft into a particular orbit such as low Earth orbit (LEO) and the means of planning planetary surface systems such as an outpost on the Moon. It also describes concept design, analysis, planning and integration of systems hardware for spacecraft in addition to explaining spacecraft/habitat layout, crew quarter design, cabin architectures and human factors analysis. This segment also describes orbital manoeuvring, orbital perturbations and orbital maintenance.

Launch Segment
The launch segment describes the launch profile of the LV from the time it leaves the launch pad to the time it enters space. It also describes vehicle-to-ground communications interface design, mission parameters and mission geometry. The launch segment also explains the process of selecting launch windows

Ground Segment
The ground segment includes all the elements required to put the spacecraft onto the launch pad and includes the process of selecting launch systems, processing facilities and operational usage assessments. Also included in this segment is an explanation of the constraints of the design process, spacecraft configuration considerations, design budgets and integration of the spacecraft design. The ground segment may also describe factors such as mission operations plans, mission operations functions and qualification programs.

Specific Impulse
In addition to thrust, rocket engineers are also interested in delivering the highest engine performance, a measure known as the *specific impulse* (I_{sp}) of a rocket. I_{sp} is measured by how many kilograms of thrust is provided by burning one kilogram of propellant in one second. Sometimes this value is difficult to calculate for different phases of a mission because the I_{sp} has a different value on the ground than in space because the ambient pressure is a factor affecting thrust, which in turn is a factor in calculating I_{sp}. Compounding the problem of accurately calculating I_{sp} are the losses occurring within the rocket engine such as the nozzle and pumps, losses which rocket engineers attempt to mitigate by designing efficient engines and nozzles.

Technology Readiness Levels
To assess the maturity of evolving technologies required to develop a new LV, NASA utilizes Technology Readiness Levels (TRLs). Some of the technologies conceptualized for the LVs and spacecraft described in this book were not suitable for immediate application. To certify the new technologies, NASA had to conduct experimentation, refinement and realistic testing until the technologies were sufficiently proven. This process followed a nine-step plan as outlined below.

NASA Technology Readiness Levels

Technology Readiness Level	Description
1. Basic principles observed and reported.	Lowest level of technology maturation. Scientific research begins to be translated into applied research.
2. Technology concept formulated.	Practical applications of technology identified. No experimental proof to support concept at this stage.
3. Analytical and experimental/ proof of concept.	Research and development initiated. Includes analytical studies to set technology into appropriate context. Studies and experiments constitute "proof-of-concept"
4. Component validation in laboratory environment.	Technological elements integrated. Ensure all the elements work together. Validation must support original concept and be consistent with requirements of potential system applications.
5. Component validation in relevant environment.	Fidelity of component increased significantly. Technological elements integrated with realistic supporting elements so total applications may be tested.
6. System/subsystem prototype demonstration.	Representative prototype system tested in a relevant environment. In many cases the only relevant environment is space.
7. System prototype demonstration in space.	Actual system prototype demonstration in space environment. Prototype should be near or at scale of planned operational system.
8. Actual system completed and 'flight qualified'.	End of system development for most technology elements. May include integration of new technology into existing system.
9. Actual system 'flight proven' through successful mission operations.	Integration of new technology into existing system. Does not include planned product improvement of ongoing or reusable systems.

Mankins, J.C. Technology Readiness Levels. A White Paper. Advanced Concepts Office. Office of Space Access and Technology. NASA. April 6, 1995.

Thrust

The propulsive force that sends the rocket skywards is termed *thrust* and is generally measured in either kilograms or Newtons. A typical rocket engine configuration features a combustion chamber with a nozzle through which the gas is expelled. The key feature to note is that pressure distribution within the chamber is not symmetrical. Inside the chamber the pressure variation is very little, but closer to the nozzle the pressure decreases. The force generated by the rocket is a function of the force due to gas pressure on the bottom of the chamber not being compensated from outside the chamber, thereby creating a pressure differential between the pressure internally and externally. This pressure differential, or thrust, is opposite to the direction of the gas jet and therefore pushes the chamber upwards.

Trans Earth Injection

Trans Earth Injection (TEI) describes the propulsive maneuver used to place a spacecraft on a trajectory that will intersect the Earth. Prior to the TEI burn the spacecraft is in a parking orbit Mars.

Trans Mars Injection

Trans Mars Injection (TMI) is a propulsive maneuver used to place a spacecraft on a trajectory that will intersect Mars. Prior to TMI burn the spacecraft is in a parking orbit around the Earth travelling at a velocity of approximately. Following the TMI burn, which is performed by a powerful rocket engine, the spacecraft will be travelling at approximately 39,400 km/h.

Index

Printing: Mercedes-Druck, Berlin
Binding: Stein+Lehmann, Berlin